AF608873

Designing Pathways for Net-Zero Greenhouse Gas Emission Plastics with Life Cycle Optimization

Wege für klimaneutrale Kunststoffe auf Basis der Lebenszyklusoptimierung

Von der Fakultät für Maschinenwesen der
Rheinisch-Westfälischen Technischen Hochschule Aachen
zur Erlangung des akademischen Grades eines Doktors
der Ingenieurwissenschaften genehmigte Dissertation

vorgelegt von

Raoul Meys

Berichter: Univ.-Prof. Dr.-Ing. André Bardow
Univ.-Prof. Dr.-Ing. Matthias Wessling

Tag der mündlichen Prüfung: 22.08.2022

Aachener Beiträge zur Technischen Thermodynamik Band 39

Raoul Meys
Designing Pathways for Net-Zero Greenhouse Gas Emission Plastics with Life Cycle Optimization
Wege für klimaneutrale Kunststoffe auf Basis der Lebenszyklusoptimierung
ISBN: 978-3-95886-463-4

Bibliografische Information der Deutschen Bibliothek
Die Deutsche Bibliothek verzeichnet diese Publikation in der Deutschen Nationalbibliografie; detaillierte bibliografische Daten sind im Internet über http://dnb.ddb.de abrufbar.

Herstellung & Vertrieb:

1. Auflage 2022

Süsterfeldstr. 83, 52072 Aachen
Tel. 0241 / 87 34 34 00
www.Verlag-Mainz.de

ISSN: 2198-4832

Satz: nach Druckvorlage des Autors
Umschlaggestaltung: Druckerei Mainz

printed in Germany
D82 (Diss. RWTH Aachen University, 2022)

"The climate crisis has already been solved. We already have all the facts and solutions. All we have to do is to wake up and change."

Greta Thunberg,
Stockholm, December (2018)

Contents

List of Figures

List of Tables

Notations

Abbreviations of chemicals and plastics

CR	chloroprene
NBR	nitrile-butadiene rubber
HNBR	hydrated nitrile-butadiene rubber
EPDM	ethylene-propylene-diene monomer rubber
cPc	cyclic polycarbonate
HDI	hexamethylene diisocyanate
HDPE	high-density polyethylene
LDPE	low-density polyethylene
LLDPE	low linear-density polyethylene
PET	polyethylene terephthalate
PS	polystyrene
GPPS	general-purpose polystyrene
HIPS	high-impact polystyrene
PP	polypropylene
NaOH	sodium hydroxide
MgO	magnesium oxide
BaO	barium oxide
CaO	calcium oxide
K2O	potassium oxide
TPA	terephthalic acid
EG	ethylene glycol
C	carbon
H	hydrogen
N	nitrogen

O	oxygen
SiC	cilicon carbide
CO_2	carbon dioxide
MeOH	methanol
NH3	ammonia

Other abbrevations

CO_2-eq	CO_2-equivalents
TCM	Technology Choice Model
CCU	carbon capture and utilization
CCS	carbon capture and storage
IEA	International Energy Agency
TRL	technology readiness level
GHG	greenhouse gas
CR	chemical recycling

Kurzfassung

Kunststoffe erobern immer mehr Bereiche des modernen menschlichen Lebens, führen aber zu einer zunehmenden Verschmutzung der Natur, einem enormen Ölverbrauch und einem massiven Ausstoß von Treibhausgasemissionen. Um den Temperaturanstieg von mehr als 1,5 Grad Celsius zu verhindern, müssen deshalb in der zweiten Hälfte dieses Jahrhunderts "net-zero" Kunststoffe ermöglicht werden. Um Treibhausgasemissionen von Kunststoffen zu reduzieren, können die neuen Kreislauftechnologien (1) chemisches oder mechanisches Recycling, (2) Kohlenstoffabscheidung und -nutzung und (3) Biomassenutzung eingesetzt werden. Die derzeitige Literatur, die diese Kreislauftechnologien bewertet, konzentriert sich jedoch auf einzelne oder nur teilweise kombinierte Kreislauftechnologien, ist regional begrenzt, oder wendet oft inkonsistente Methoden an. Daher ist es derzeit unklar, ob "net-zero" Emissionen bei Kunststoffen mit den neuen Kreislauftechnologien tatsächlich erreicht werden können.

Um die Möglichkeit von "net-zero" Kunststoffen zu analysieren, wird in dieser Arbeit das erste globale, branchenweite und systematische "bottom-up"Modell für die Produktion und die Abfallbehandlung von Kunststoffen entwickelt und verwendet. Dieses Modell deckt die globalen Treibhausgasemissionen von 90% der weltweiten Kunststoffe ab. Die Anwendung zeigt, dass Kunststoffe mit "net-zero" Emissionen durch die Kombination von Biomasse- und CO_2-Nutzung mit einer effektiven Recyclingrate von 70% erreicht werden können, während bis zu 53% Energie und 288 Milliarden USD im Vergleich zu einem fossilen Benchmark mit groß angelegter Kohlenstoffabscheidung und -speicherung eingespart werden können. Um das volle Potenzial an Energie- und Kosteneinsparungen auszuschöpfen und gleichzeitig "net-zero" Emissionen zu erreichen, muss die Versorgung mit Biomasse und CO_2 zu niedrigen Kosten erfolgen, während die Kosten für die Ölförderung- und bereitstellung erhöht werden müssen. Dazu müssen die Investitionsbarrieren für alle verfügbaren Kreislauftechnologien gesenkt werden, indem konsequente Emissionspreisregelungen eingeführt werden, Pfandsysteme für Kunststoffe gefördert werden, und die Subventionierung fossiler Ressourcen eingestellt wird. Diese Arbeit zeigt also, dass das Problem der Treibhausgasemissionen von Kunststoffen mit bereits heute verfügbaren Technologien und Ansätzen gelöst werden kann.

Abstract

Plastics are on the rise to conquer every area of modern human life but lead to increased pollution of nature, enormous oil consumption, and large-scale greenhouse gas emissions. Thus, to avoid climate change above 1.5°C, net-zero greenhouse gas emission plastics are needed by the second half of this century. To reduce the greenhouse gas emissions associated with plastics, three circular technologies can be used: (1) chemical or mechanical recycling, (2) carbon capture and utilization, and (3) biomass utilization. However, current environmental assessments of these circular technologies focus solely on individual or partly combined circular technologies, are limited to regional scopes, and often apply inconsistent methodologies. Thus, it is currently unclear if net-zero emission plastics can actually be achieved with the current set of circular technologies. Furthermore, shifting from the linear to a circular economy is regarded as energy-intensive and costly, hindering strong policy implementation from fostering the transition to a circular economy.

To assess if net-zero emission plastics can actually be achieved, this thesis builds and uses the first global, industry-wide and systematic bottom-up model for plastics production and waste treatment, representing the global life cycle greenhouse gas emissions of 90% of global plastic production. Using that model reveals that net-zero emission plastics can be achieved by combining biomass and CO_2 utilization with an effective recycling rate of 70% while saving up to to 53% of energy and 288 billion USD compared to a fossil-based benchmark applying large-scale carbon capture and storage. Achieving the full potential of energy and cost savings while achieving net-zero emissions requires the supply of biomass and CO_2 at low cost, while cost of oil supply must be increased. To incentivize this shift, investment barriers for all available circular technologies have to be lowered by implementing consistent emission pricing schemes, using deposit systems for plastics to increase recyclability and stopping to subsidize fossil resources. Thus, this thesis shows that the greenhouse gas emission problem of plastics can be solved with technologies and solutions already available today.

CHAPTER 1

Introduction

Since the early 1950s, synthetic plastics have made their way into nearly every area of modern life. Plastics are used in almost every sector, from packaging, transportation, buildings, and electronics to healthcare. They provide substantial economic benefits for these sectors by combining low costs and enhanced product performance. Examples of enhanced product performance are increased food durability or decreased fuel consumption due to lower weight-to-strength ratios (OECD, 2018; Andrady and Neal, 2009). As a result, plastic consumption 20-folded between 1964 and 2014 from 15 to 311 Mt per year. Following this exponential growth, plastic production is expected to increase to 1124 Mt annually in 2050 (Foundation, 2016). While plastics' economic and practical attractiveness are evident, their environmental impacts gained increasing attention from policy, industry, science, and the public.

This attention is based on the immense amount of plastic waste in the maritime and terrestrial natural environment (Haward, 2018; Isensee and Valdes, 2015). In fact, plastics are seen as the epitome of the "take-make-dispose" or "linear" economy since *"of all plastics ever made, only 9% have been recycled globally"* (Geyer et al., 2017) and *"the ocean is expected to contain 1 tonne of plastic for every 3 tonnes of fish by 2025, and by 2050, more plastics than fish"* (Foundation, 2016). While plastic waste historically has been at the center of environmental concerns, another major ecological threat recently appeared in the public perception: the impact of plastics on the global climate (Zheng and Suh, 2019). Currently, the production of plastics consumes over 80% of basic chemicals like ethylene and propylene, which are exclusively based on fossil resources, particularly crude oil (Levi and Cullen, 2018). In return, plastics and their chemical raw materials are expected to be responsible for 20% of the global crude oil consumption by 2050 (IEA, 2018a). This extensive use of fossil-based resources for plastics led to 1.8 Gt of CO_2-eq emissions in 2015 (Zheng and Suh, 2019). Continuing how humanity produces plastics and treats plastic waste, the plastic life cycle greenhouse gas emissions are even expected to claim 15% of the yearly allowed greenhouse gas emissions by 2050 to keep global warming below 1.5°C (Foundation, 2016). Thus, capping temperature rise to 1.5°C will unavoidably require net-zero emission plastics in the second half of this century (Rogelj et al., 2018).

Measures to achieve net-zero emission plastics include the decarbonization of energy supply and the implementation of circular technologies (Bazzanella and Ausfelder, 2017; Zimmerman et al., 2020) such as (1) chemical or mechanical recycling, (2) carbon capture and utilization, and (3) biomass utilization. By recycling all plastic packaging materials back to their original raw material via mechanical recycling, reductions of 25% of plastics' climate impacts have been estimated (Zheng and Suh, 2019). Reducing carbon emissions by mechanical recycling is limited due to material losses that ultimately must be incinerated (Ragaert et al., 2017). For carbon capture and utilization, individual technologies have been reported to reduce greenhouse gas emissions (Artz et al., 2018). Current estimates for the global potential of carbon capture and utilization to reduce greenhouse gas emissions, however, rely on the amount of CO_2 used in the respective products (Mac Dowell et al., 2017). But, for instance, using 1 kg of CO_2 in polyurethanes can avoid up to 3 kg of CO_2 (von der Assen and Bardow, 2014). Thus, the amount of used CO_2 is not sufficient to estimate the global potential of carbon capture and utilization to reduce greenhouse gas emissions (Mac Dowell et al., 2017; DG RTD, 2018b,a). By utilizing biomass, climate benefits of up to 1 Gt CO_2-eq emission per year were estimated in 2007 if only some chemicals and plastics are taken into account (Hermann et al., 2007).

Thus, the literature focuses on the assessment of individual or partly combined circular technologies and the respective potential for large-scale reductions of greenhouse gas emissions (Hermann et al., 2007; Artz et al., 2018; Zheng and Suh, 2019; Schwarz et al., 2021; Posen et al., 2017; Saygin et al., 2014). However, individual studies often differ in underlying assumptions, data, or methodological choices and are, thus, not directly comparable (Laurent et al., 2014a). Furthermore, no study exists for plastics that (1) systematically evaluates all circular technologies simultaneously and (2) assesses if the circular technologies can actually achieve net-zero greenhouse gas emissions on a global, industry-wide scale. As a result, a comparable, systematic, and industry-wide assessment of circular technologies to enable net-zero emission plastics is currently missing.

To close this gap, scientists face three major challenges: First, a large amount of novel circular technologies have to be evaluated against their conventional benchmarks, despite low data availability. Due to this low data availability, making robust decisions for early-stage technologies and identifying the most promising technologies is difficult. Second, even though many mature circular technologies exist, data about their environmental impacts and potential benefits have to be collected. Gathering this data is time-consuming and costly, which is why the data is not available in a consistent manner. Finally, all data about early-stage and mature technologies would have to be combined in a consistent and industry-wide model to systematically evaluate if and how net-zero emission plastics can actually be achieved.

To overcome these three challenge, this thesis builds and uses the first global and industry-wide bottom-up model for plastics production and waste treatment, representing the global life cycle greenhouse gas emissions of 90% of global plastic production. This bottom-up model is based on the methodology of the Technology Choice Model (Kätelhön et al., 2016) and utilizes over 400 technology datasets that are derived based on the life cycle assessment standards (ISO 14040, 2021; ISO 14044, 2021). The model applies a consistent methodology for all datasets to avoid ambiguity due to methodology. The bottom-up model is then used to systematically highlight the potential reductions of greenhouse gas emissions by recycling plastic waste and using the renewable carbon sources CO_2 and biomass. Thus, this thesis assesses technically feasible combinations of all circular technologies to achieve net-zero emission plastics. Furthermore, this thesis calculates the energy consumption and the respective operational costs of these net-zero emission plastic life cycles. By this means, this thesis shows that, under favorable market conditions, net-zero emission plastics are achievable with lower energy demands and lower operational costs than fossil-based production technologies combined with large-scale carbon capture and storage.

1.1 Structure of this thesis

The following Chapter 2 provides an overview of the current status of the life cycle assessment of plastics and their chemical raw materials. First, a technology perspective on the changing life cycle of plastics from fossil and linear to renewable and circular plastics is presented. This overview includes the major changes in feedstock sources as well as production and waste treatment technologies. By this means, the chapter sets the scene about the expected shifts in industry and highlights which circular technologies need to be included in the thesis scope. Second, the current status of life cycle assessment of net-zero emission pathways towards plastic life cycles is outlined based on available scientific and public literature. This literature review aims to identify the components of a life cycle assessment to evaluate the transition to net-zero emission plastics. Based on the literature review about technologies and life cycle assessment, the major scientific gaps in literature and contributions of this thesis are summarized.

Chapter 3 addresses the environmental assessment of chemical recycling technologies that are currently in early-development stages. Due to this early-development stage, the environmental impacts of these novel chemical recycling technologies are not fully understood. Thus, Chapter 3 evaluates chemical recycling technologies from an environmental perspective and proposes a methodology for robust environmental decisions already at early-development stages.

In Chapter 4, the life cycle assessment standard is applied to the plastic case study of CO_2-based rubbers. By this means, it is highlighted that a life cycle perspective including plastic production and end-of-life is needed to develop sound conclusions about environmental benefits or disadvantages of CO_2-based products. Thus, the study highlights that datasets to perform life cycle assessments of circular technologies need to represent both production and end-of-life of plastic materials. The knowledge about these components and datasets build the basis for the subsequent chapters that present an industry-wide perspective on plastic products.

Chapter 5 addresses the methodological and data-related developments performed during this thesis. In particular, the bottom-up model of plastic life cycles is presented. The bottom-up model represents the supply chain of chemicals and plastics production and end-of-life technologies based on over 400 detailed datasets. By this means, the bottom-up model enables a consistent, industry-wide and systematic assessment of all circular technologies to achieve net-zero emission plastics.

Based on the bottom-up model, Chapter 6 presents results for the fossil and linear plastic life cycle for the years 2015 and 2050. The results for 2015 are used to validate the bottom-up model results with current literature values, while the results for 2050 are used as a reference in the subsequent Chapter 7.

Chapter 7 uses the model to show pathways for net-zero emission plastics and highlights potential policies to foster the transition towards the net-zero future. In particular, Chapter 7 shows that the optimal combination of circular technologies achieves net-zero emission plastics while potentially requiring less energy and operational costs than a fossil and linear life cycle that achieves net-zero emissions through large-scale carbon capture and storage.

Chapter 8 summarizes the conclusions from this thesis, discusses assumptions and limitations and provides an outlook for future work in the field of environmental assessments of plastic life cycles.

Chapter 2

State of the art life cycle assessment of plastics

This chapter provides a general overview of recycling, CO_2 utilization, and biomass utilization (together denoted as circular technologies) as well as their respective potential to reduce greenhouse gas emissions.

However, before providing a more detailed discussion about the circular technologies, the term "net-zero emission" plastics is defined and distinguished from the terms of "carbon-neutral", "carbon-negative", "net-negative emission", and "zero-emissions" plastics (Section 2.1). The definition is important to avoid misunderstanding the terminology used in this thesis.

After defining net-zero emission plastics, Section 2.2 compares today's linear (Section 2.2.1) to tomorrow's circular life cycle of plastics (Section 2.2.2). To outline the future developments, the three circular technologies (1) plastic waste recycling, (2) carbon capture and utilization, and (3) biomass utilization are discussed. By this means, the section sets the scene about the expected shifts in industry and highlights those circular technologies that need to be included in the bottom-up model developed in this thesis (Chapter 5).

Section 2.3 provides an overview of how life cycle assessment is used to assess the three circular technologies. Based on a short description of the fundamentals of life cycle assessment in Section 2.3.1, its practical application to individual circular technologies is summarized in Section 2.3.2. Furthermore, Section 2.3.3 provides an overview of literature assessing the industry-wide potential of circular technologies to reduce greenhouse gas emissions of plastics.

Finally, the scientific gaps identified during in this thesis are summarized in Section 2.4. The final Section 2.5 highlights how this thesis addresses the scientific gaps in the subsequent Chapters 3 to 7.

2.1 Carbon-neutral, net-zero, carbon-negative, or net-negative plastics?

Many different terms are used in literature that describes the greenhouse gas emissions or the climate impact of products and economies (Tanzer and Ramírez, 2019; Gabrielli et al., 2020). These terms range from carbon-neutral and net-zero emissions, over carbon-negative and net-negative emissions to zero-emission and are used interchangeably, introducing uncertainty to their actual meaning. To avoid misunderstandings throughout this thesis, the used terms and their meaning are defined in the following.

Net-zero emission products, companies, or even whole economies do not change the greenhouse gas emission concentration in the atmosphere since the same amount of greenhouse gas emissions is emitted as are absorbed or sequestered (IPCC, 2021). As a result, human activities have no effect on the global climate.

In contrast, the term *carbon-neutral* or net-zero CO_2 only include CO_2 emissions according to the glossary of the International Panel of Climate Change. Thus, the term carbon-neutral does not include other greenhouse gases like nitrous oxides or methane.

In principle, plastic material can achieve net-zero emissions in two ways (Figure 2.1): The first way is when greenhouse gas emissions are absorbed from the atmosphere and later re-emitted. The second possibility is the usage of fossil-based resources in combination with subsequent CO_2-sequestration of an equivalent amount of greenhouse gases emitted during the complete life cycle (IPCC, 2021). To evaluate whether a plastic material actually achieves net-zero emissions, the assessment needs to (1) consider the full life cycle of products, e.g., from cradle-to-grave and (2) include all greenhouse gas emissions and should not focus solely on CO_2 (Finkbeiner and Bach, 2021).

Similar to the carbon-neutral and net-zero emissions, the terms *carbon-negative* and *net-negative emissions* correspond to either considering only CO_2 or all greenhouse gases, respectively. Plastic materials can be carbon-negative if they sequester more CO_2 or net-zero emission if they sequester more greenhouse gases than are emitted over the complete life cycle by (1) physically removing greenhouse gas emissions from the atmosphere and (2) storing these greenhouse gases in a permanent manner. Here, in particular, the inclusion of credits or offsets poses a risk for counterproductive misunderstanding in decision-making (Tanzer and Ramírez, 2019).

In contrast to carbon-neutral, net-zero emissions, carbon-negative, or net-negative emissions, zero-emission human activities, including plastic production and usage, are

Figure 2.1: Two potential ways of reaching net-zero emission or carbon-neutral plastics.

pure fiction from a life cycle perspective. The "zero-emission" terminology implies that there are no greenhouse gas or other emissions at all over the complete life cycle. One frequent example referred to as "zero-emission" is the electric vehicle. In fact, no emissions occur at the (non-existing) exhaust, but the production of electricity leads to greenhouse gas emissions if produced from fossil-based resources and even if produced from renewable resources. For instance, producing biomass for bio-based electricity emits greenhouse gases like methane or nitrous oxides. Even wind-based electricity leads to greenhouse gas emissions since the infrastructure, for instance, requires cement for the foundations and cement production leads to stoichiometric CO_2 emissions.

As a result of the aforementioned discussion and to avoid confusion in this thesis, only the term net-zero is used since all greenhouse gases emissions are included in the subsequent calculations.

2.2 Comparing today's linear with tomorrow's circular life cycle of plastics

To keep global temperature rise below 1.5°C, anthropogenic greenhouse gas emissions of all industries must turn net-zero by 2050 (Rogelj et al., 2018). In contrast to other industries, the petrochemical and plastic industry faces a two-fold challenge to achieve net-zero emission life cycles since it needs to eliminate greenhouse gas emissions from (1) process energy and (2) the utilization of fossil carbon as the feedstock of plastics.

Today, the plastic life cycle is almost completely running on fossil-based carbon feedstock, originating from refinery operations (Figure 2.2, top). In these refinery operations, crude petroleum is upgraded to fuels and chemical feedstock, like naphtha or liquid petroleum gas. The chemical feedstock is then converted to chemical monomers. These chemical monomers are then utilized to produce the plastic products we currently consume (Elvers and Ullmann, 2011). At their end-of-life, these plastics are then frequently incinerated in waste incinerators leading to further greenhouse gas emissions or are disposed in landfills, leading to a loss of the incorporated carbon feedstock (Geyer et al., 2017).

Tomorrow, this linear "make-use-dispose" life cycle of plastics is about to change circular (Figure 2.3, bottom). This circular life cycle of plastics is running based on renewable energy from wind or biomass and, to fully turn net-zero, circulates carbon. The carbon cycles are based on three key circular technologies (Zimmerman et al., 2020): (1) mechanical and chemical recycling of plastic waste (2) carbon capture and utilization, and (3) biomass utilization.

2.2.1 Today: fossil and linear

Since the plastic industry is based on chemical monomers that are based on chemical feedstock from refinery operations, the plastic industry is inevitably connected to the usage of fossil resources in these refineries (Figure 2.2, top). In fact, the chemical industry is the single largest industrial consumer of fossil energy, ahead of iron and steel as well as cement (IEA, 2018a). The chemical industry represents 10% of the global and almost 30% of the industrial energy demand. This energy demand is driven by the utilization of oil, gas, and coal to provide process energy and carbon feedstock. The following paragraphs will provide a high-level view of the current supply chain of plastics from fossil raw materials over plastic use to their end-of-life. By this means, this section enables the identification of possibilities to integrate circular technologies (Section 2.2.2) inside the current life cycle of plastics.

In 2015, the share of fossil feedstock equaled approximately 58% of the overall energy input to the chemical industry. This feedstock can be subdivided into oil (80%), natural gas (13%), and coal (7%) (IEA, 2018a)[1]. These shares are representative of the production of the major chemicals ammonia and methanol, as well as ethylene,

[1]This calculations assumed 12 million barrels per day of oil, 105 billion cubic meters of natural gas and 80 Mt of coal (IEA, 2018a). To convert these values to MJ per year, 136 kg per barrel, 41.868 MJ per barrel, 38.2 PJ per billion cubic meter of natural gas, and 29.3076 GJ per ton of coal were used as conversion factors.

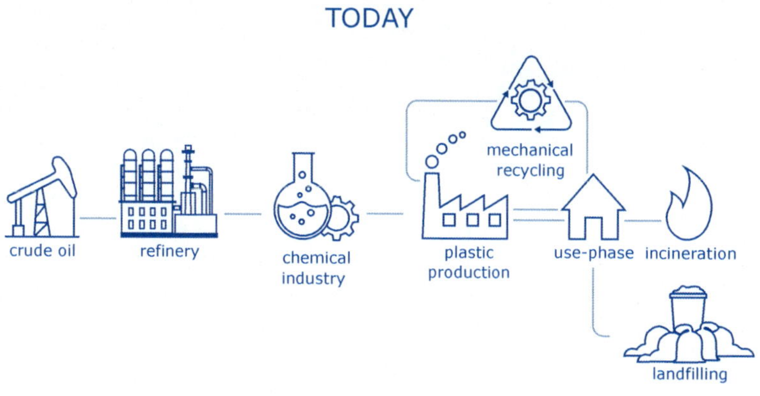

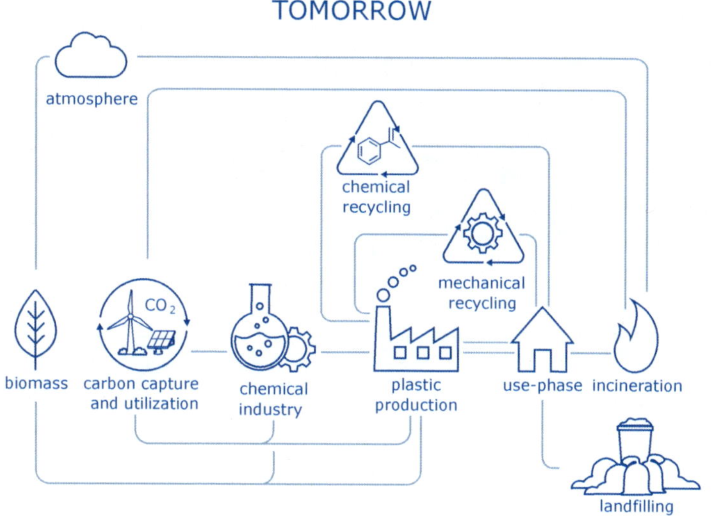

Figure 2.2: The changed life cycle of plastics from a linear to a circular economy.

propylene, benzene, toluene and xylene (the last five are denoted together as high-volume chemicals). Ammonia and methanol utilize almost exclusively natural gas (72% and 49%) or coal (28% and 50%) as carbon feedstock (Figure 2.3) (IEA, 2018a). While natural gas utilization for ammonia and methanol is distributed equally over countries, coal is exclusively used in China and India (ICIS, 2018; IEA, 2018a). In the case of methanol production, a relatively small share of feedstock is based on heavy oil gasification, which is mainly practiced in Germany. In contrast, 96% of the high-volume chemicals use crude oil as feedstock. The crude oil feedstock is treated in refineries to produce primarily liquid hydrocarbons like naphtha or gas oil that are then converted to high-volume chemicals via steam cracking with subsequent separation of aromatics via extraction, crystallization and adsorption (Elvers and Ullmann, 2011). Coal and natural gas only play a minor role, with 4% globally. These 4% coal-based production are due to the large-scale production of ethylene and propylene from methanol and the extraction of benzene, toluene and xylenes from crude benzole (a coke-oven by-product) in China (IEA, 2018a).

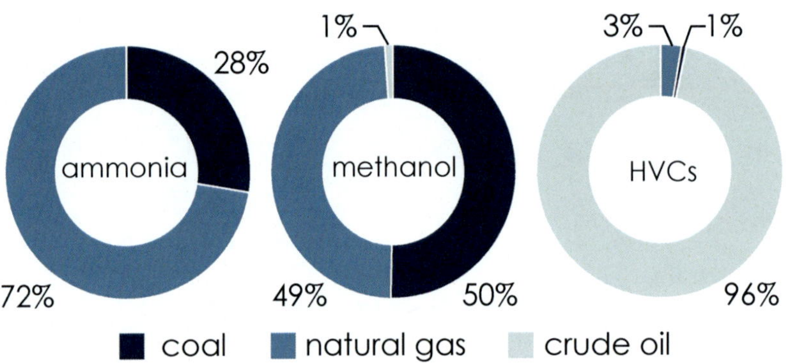

Figure 2.3: Share of fossil feedstock for ammonia, methanol and high-volume chemicals (HVCs: ethylene, propylene, benzene, toluene, xylenes). Crude oil is mostly treated by refineries to produce liquid hydrocarbons like naphtha that build the major chemical feedstock (IEA, 2018a).

Up to 80% of ammonia is used in the agricultural sector and, thus, ammonia plays only a minor role in the plastics industry (IEA, 2018a). Methanol and high-volume chemicals build the major chemical raw materials for plastics. In 2018, methanol was also used for fuels and other chemical products (together 63%) that are not used in plastics (ICIS, 2018). Methanol utilization for plastics production is mostly performed via methanol to high-volume chemical technologies (e.g., methanol to ethylene/propylene), representing 28% of the overall methanol utilization. Smaller shares of methanol are converted to methyl methacrylate and acetic acid (together 9%), which are also partly (approx. 50%) used in polymers but play a minor role in methanol utilization today. Thus, the focus of the following paragraphs is on the conversion of high-volume chemicals to plastics.

Of each high-volume chemical, over 80% has been used in plastics in the year 2018 (Figure 2.4). Note that the shares in the following have been calculated based on data included in ICIS (2018) and IEA (2018a). For the olefins ethylene and propylene, the largest share has been used for the direct polymerization to polyethylene (60%) or polypropylene (66%). Besides the direct polymerization, ethylene and propylene are oxygenated to ethylene oxide and then further to ethylene glycol, or propylene oxide, respectively. These oxygenates are then used for the production of polyesters and polyurethanes. For ethylene, 9.4% was used for polyesters, particularly polyethylene terephthalate, in 2018. For propylene, polyurethanes via propylene oxide represented 3.6% of propylene usage. In addition, 9.5 % of ethylene is also chlorinated to vinyl chloride which is then polymerized to polyvinyl chloride and 3.3% of ethylene are used for the alkylation of benzene to produce styrene for polystyrene production.

The aromatic products benzene, toluene, and xylenes are most commonly used to form either polystyrene (19.0%) or polyesters (43.4%). Here, two major intermediates are used: (1) styrene produced via the alkylation of benzene with ethylene plus the subsequent dehydrogenation of ethylbenzene and (2) terephthalic acid produced by the oxidation of para-xylene. While styrene is polymerized directly to polystyrene, terephthalic acid is used in a polycondensation reaction together with ethylene glycol to form polyethylene terephthalate, the most common polyester. Additionally, benzene can be alkylated with propylene to produce cumene which is then oxidated to produce phenol and acetone. Phenol is used in many applications, of which bisphenol A represents the largest share. Polycarbonate from bisphenol A represents 3.1% of the aromatics utilization share. Finally, aromatic components are used for polyurethanes (5.3%) and polyamides (1.1%). For polyurethanes, benzene is nitrated to nitrobenzene, which is subsequently hydrogenated to aniline, the major precursor for methylene diphenol diisocyanate used in polyurethanes. To produce polyamides, caprolactam is produced from cyclohexanone which can be produced by the partial hydrogenation of phenol or the oxidation of cyclohexane.

Finally, other applications exist for ethylene (4.8%), propylene (10,4%), and benzene, toluene and xylenes (14.2%). These other plastics comprise various products, such as rubbers (e.g., styrene-butadiene rubbers), elastomers for electronic products (e.g., acrylonitrile-butadiene-styrene or ethylene-propylene-norbornene resins), but also adhesives (e.g., epoxy resins) or high-performance plastics (e.g., polysulfones, polyetheretherketones or polyetherimides) (Wypych, 2012). Since the supply chains of most other plastics are more complex, they is not shown in Figure 2.4 to increase readability and to focus on the general supply chain of the major plastic materials.

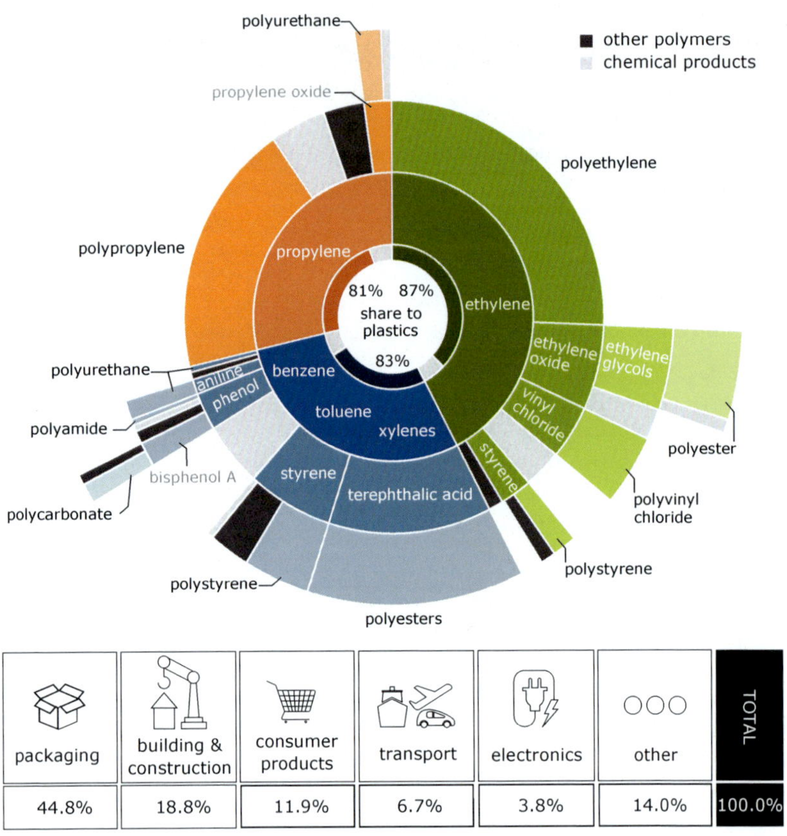

Figure 2.4: Overview of the share of ethylene, propylene, benzene, toluene and xylene utilization for plastics applications (ICIS, 2018) and the most important sectors for plastics (Geyer et al., 2017).

The plastics derived from the high-volume chemicals are used in diverse industrial sectors like building and construction (18.8%), consumer products (11.9%), transportation (6.7%), electronics (3.8%), and several other uses that sum to 14% (Figure 2.4). However, the largest single use of plastics with 44.8% is the packaging industry (Geyer et al., 2017). While plastic packaging is important, for instance, to increase food durability, its intensive usage has also resulted in an increased awareness of the global plastic waste problem since 61% of the beach litter is based on packaging materials (Haward, 2018; Isensee and Valdes, 2015). Plastic debris is present in all major ocean basins in the world, including remote islands, the poles, and deep seas (Barnes et al., 2009). This plastic debris is almost impossible to handle for the natural environment since its inert character prohibits natural degradation mechanisms. As a result, plastic debris in the natural environment would take several hundreds of years to degrade (Chamas et al., 2020). Even worse, microscopic plastic fragments leak into the maritime waters and are ingested by maritime life. Through sea food, these microplastics even reach the human metabolism and, thus, damage not only the ecosystem but also human health. The reason for these damages is simple: the current way of managing plastic waste (Foundation, 2016).

Plastic waste management is dominated by the public sector, in which plastic waste is collected as part of a mixed waste stream or, for instance, in parts of Europe, as a single waste fraction (OECD, 2018). Depending on the system, plastic waste is treated either separately or as part of the municipal solid waste system. In general, three options for waste treatment options exist (Singh et al., 2017): plastic waste can (1) be *disposed* in managed systems (e.g., landfills) or open dump sites, e.g., the natural environment, (2) be *incinerated* with or without energy recovery (heat and/or electricity), and (3) be *recycled* by mechanical or chemical recycling technologies.

Until 1980, plastic waste incineration and recycling was not practiced and most of the plastic waste was disposed. As a result, of the 6.3 billion tons of plastic waste produced over time, only 9% have been recycled and only 0.9% were recycled more than once. Of all plastics ever made, 11% have been incinerated and 80% have been landfilled or dumped in the natural environment (Geyer et al., 2017).

To cope with the ever increasing amount of plastic waste, waste management schemes have been developed in high-income countries, particularly Europe, since 1980. Global recycling rates increased to 14-18% and incineration rates increased to 24% by 2015. However, still, 58-62% of the yearly plastic waste is disposed in landfills or dumped into the environment (Geyer et al., 2017; OECD, 2018). Recently, the annual amount of plastic waste leaked to the natural environment was estimated to be about 80 Mt in 2015 (Lebreton and Andrady, 2019), being equal to approximately half of the global plastic waste generation. The exponential growth of plastic waste has been a particular challenge in regions with rapid economic development and population growth (Brooks et al., 2018). Thus, the global waste treatment rates do not properly represent regional trends. While comprehensive region- and polymer-specific data is missing, an overview of waste management rates of China, the USA and Europe, as well as the global average is provided in Figure 2.5. Currently, Europe and China achieve recycling rates of 31% and 25%, respectively (Geyer et al., 2017; PlasticsEurope, 2018). In contrast, the recycling rates in the USA have remained constantly low, around 9%, since 2012 (U.S. Environmental Protection Agency, 2020). Incineration rates increased to 42% and 30% in Europe and China, respectively. Landfill rates dropped in Europe and China, but rates are constant at around 75% in the USA.

Thus on a global picture, the effort to increase recycling rates is increasing in many, but not all regions. However, recycling rates of plastics are still below those of other materials such as paper (58%) or iron and steel (70-90%) (van Ewijk et al., 2018). As a result of the low recycling rate, plastics have become the epitome of the "take-make-dispose" or "linear" economy.

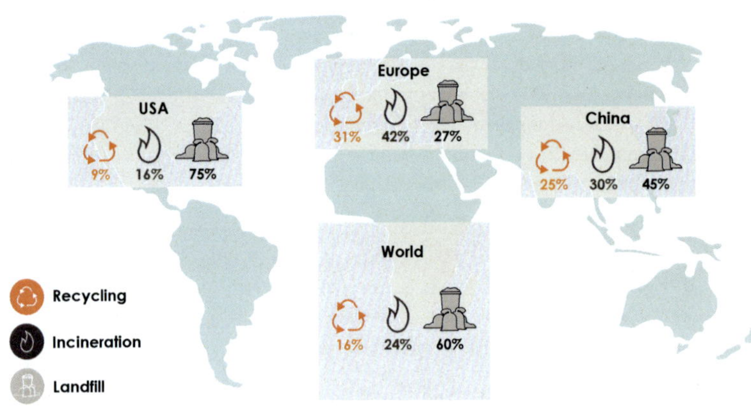

Figure 2.5: The share of waste treatment technologies in Europe, the USA, China and the global average (Geyer et al., 2017; PlasticsEurope, 2018; U.S. Environmental Protection Agency, 2020).

2.2.2 Tomorrow: renewable and circular

Based on the exponential growth of plastics and the devastating plastic waste management, linear and fossil-based plastics led to an ever-increasing amount of plastic pollution, greenhouse gas emissions, and fossil resource depletion (Lazarevic et al., 2010). One option to reduce the environmental impacts of plastics is not producing plastics in the first place. While this management of demand will play a crucial role, producing no plastics at all seems unrealistic due to the economic- and performance-related benefits plastic products offer (OECD, 2018; Andrady and Neal, 2009). Thus, this thesis will focus on reducing environmental impacts by technological measures to (1) reduce fossil resource utilization to provide process energy and (2) close the carbon loop. Fossil resources used for process energy can be replaced with renewable energy, e.g., coal-based electricity with biomass-based electricity. Furthermore, efficiency improvements or completely new process concepts can decrease the overall energy demand. Closing the carbon loop requires disruptive changes by the large-scale implementation of three types of circular technologies (Zimmerman et al., 2020): (1) chemical and mechanical recycling, (2) carbon capture and utilization, and (3) biomass utilization. Furthermore, carbon capture and storage can store CO_2 emissions underground (Gabrielli et al., 2020). In the following paragraphs, this thesis highlights the major possibilities to integrate circular technologies into the current life cycle of plastics. Furthermore, this section provides an technical overview of already industrialized circular technologies, e.g., at pilot or industrial scale, as well as an outlook about future, early-development, options.

All three circular technologies offer a multitude of process concepts, each of which includes several variations of apparatuses, process conditions, or catalyst systems. However, this vast amount of configurations can be narrowed down to only a few generalized options for integration in the current supply chain of plastics (Figure 2.6): The first option is the production of chemical feedstock like liquid hydrocarbons, e.g., naphtha and gas oils, and synthesis gas, a mixture of carbon monoxide and hydrogen. This chemical feedstock currently builds the basis for the production of high-volume chemicals (ethylene, propylene, benzene, toluene, and xylene) and methanol. The second option is the direct production of methanol, high-volume chemicals, and other plastic intermediates from renewable resources. Producing chemical feedstock, high-volume chemicals, methanol, or plastic intermediates offers the potential to keep already existing infrastructure since chemically identical molecules to the current fossil-based pathways are produced. Finally, circular technologies can be used to directly produce the existing plastics, e.g., via recycling, or novel polymers, e.g., polylactic acid or polyhydroxyalkanoates. In these cases, completely new supply chains must be established.

The following paragraphs describe the circular technologies in more detail.

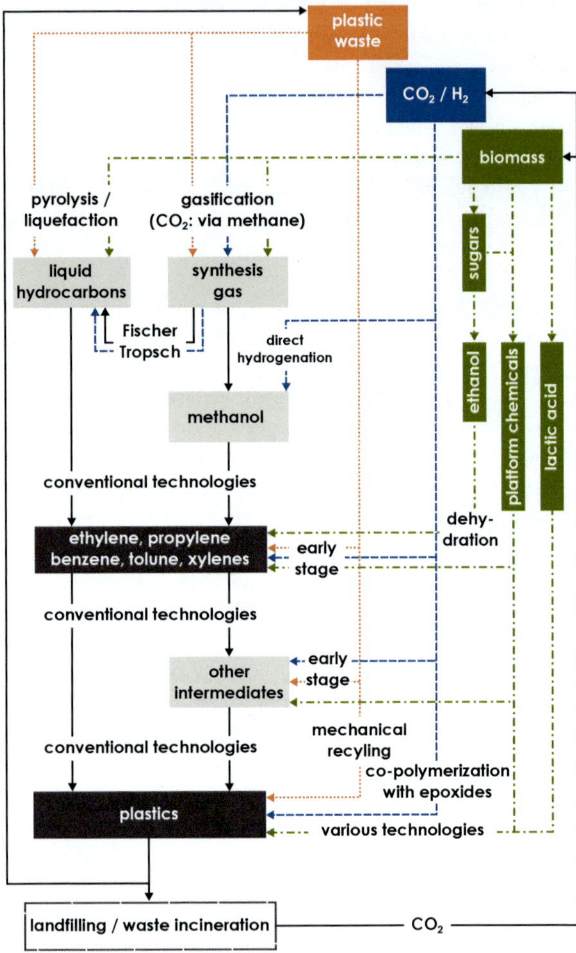

Figure 2.6: Potential circular technologies to produce plastics from the renewable resources plastic waste, CO_2 and H_2, and biomass.

Recycling of plastic waste

The first option to circulate carbon is based on the utilization of plastic waste via mechanical and chemical recycling (Figure 2.6 dotted / orange lines). For both technologies, the possibility for integration into existing plastic supply chains differs.

Mechanical recycling leads to the recovery of plastic resins that can be re-used in manufacturing processes for new plastic materials (Al-Salem et al., 2009). Thus, mechanical recycling substitutes the complete supply chain from fossil resources to plastic materials. However, mechanical recycling requires new supply chains from plastic waste to new plastic re-granulates. To produce re-granulates, mechanical recycling requires the first step of plastic waste sorting. Typically, plastics are sorted by shape, density, size, or chemical composition. Afterward, the plastic waste is washed and crushed into smaller pieces that are extruded into new plastic pellets or granulates. These mechanical treatments frequently lead to a loss of material properties, so-called "downcycling" (Ragaert et al., 2017).

In contrast, chemical recycling technologies convert plastics into chemicals (Al-Salem et al., 2009; Ragaert et al., 2017). Currently, a major effort in chemical recycling is in the recovery of liquid or gaseous hydrocarbon products by using pyrolysis or gasification technologies, respectively (Ragaert et al., 2017). During pyrolysis, plastic waste is broken down into smaller carbon chains, leading to, primarily, liquid hydrocarbon products. To increase yield to liquid products, pyrolysis technologies utilize short residence times, ambient pressures, and temperatures between 400 and 600 °C. These liquid hydrocarbon products can be used in the existing chemical production technologies, e.g., steam cracking, to produce all high-volume chemicals. Currently, several companies, for instance, Quantafuel or Plastic Energy, have developed industrialized pyrolysis processes. In fact, large chemical producers like BASF, LyondellBasell, or Eastman are heavily investing in pyrolysis technologies in order to secure the future supply of plastic-derived steam cracking feedstock.

Gasifying plastic waste produces synthesis gas at temperatures above 1000 °C (Ragaert et al., 2017; Schwarz et al., 2021). The plastic waste-derived synthesis gas is then enriched with hydrogen via a water-gas-shift reaction to meet the required carbon-monoxide-to-hydrogen ratio. Plastic waste gasification is already developed to commercial scale, for instance, by Enerkem. Based on the synthesis gas, for instance, methanol can be produced and utilized for plastic production through existing supply chains (Section 2.2.1).

Besides the production of liquid hydrocarbons and synthesis gas, chemical recycling technologies can also be used to directly recover high-volume chemicals and other plastic monomers (e.g., other intermediates), like terephthalic acid. However, chemical recycling technologies to produce the plastics monomers directly are in early-development stages or pilot-scale and are not yet industrial standard (Clark et al., 2016; Rahimi and García, 2017; Hong and Chen, 2017). Some exceptions exist for the chemical recycling of polyethylene terephthalate. For instance, terephthalic acid and ethylene glycol are recovered from polyethylene terephthalate production residues (Paszun and Spychaj, 1997) and polyethylene terephthalate bottles are upcycled chemically to produce polyol precursors for polyurethanes by Huntsman (McCoy, 2020).

Carbon capture and utilization

The second carbon loop can be established by carbon capture and utilization technologies that employ CO_2 as raw material (Figure 2.6 dashed / blue lines). CO_2 can be extracted from exhaust gases of point sources or from ambient air (von der Assen et al., 2014). To activate the inert molecule CO_2, a high-energetic chemical co-reagent is usually needed. While epoxides have been used to produce some polyurethane products at an industrial scale by Covestro (Langanke et al., 2014), the most frequent co-reagent for CO_2 is H_2.

Currently, two major intermediate CO_2-based products are based on CO_2 and H_2: methane and methanol. For both intermediates, already commercialized processes exit. Methanol is produced by Carbon Recycling International, for instance, in a plant in Iceland, while methane is produced by, for instance, Uniper. Methanol is produced directly from CO_2 and H_2 via direct thermal hydrogenation. The direct hydrogenation is based on copper and/or aluminum oxide catalysts and operates at about 200°C and 65 bar (Artz et al., 2018). Producing methane from CO_2 and H_2 is based on the Sabatier reaction. The Sabatier reaction is already commercial practice to purify H_2 during ammonia synthesis (Navarro et al., 2018). The CO_2-based methane can be converted to synthesis gas and subsequently used to produce liquid hydrocarbons or methanol. Liquid hydrocarbons are already industrially produced via the commercialized Fischer-Tropsch synthesis by, for instance, Sasol in South Africa or Shell in Qatar. Furthermore, synthesis gas can be used to produce methanol by existing technologies.

Besides the production of methane and methanol, several early-stage technologies exist to directly produce the chemical monomers benzene, toluene and xylenes (Wang et al., 2014b), as well as ethylene and propylene (Gao et al., 2017; Yabe et al., 2017; Yang et al., 2017). Furthermore, a scientific effort is directed to the electrochemical production of various hydrocarbon products like ethylene or methanol (Dinh et al.,

2018; de Luna et al., 2019; Kriescher et al., 2015).

Biomass utilization

The third carbon loop is based on biomass utilization (Figure 2.6 dotted-dashed / green lines). The biomass feedstock can, in general, be divided into (1) sugary and starchy crops including corn or sugarcane, (2) lignocellulosic biomass such as miscanthus or wood residues, and (3) oily biomass like rapeseed or soybeans (Muñoz et al., 2014; Gerssen-Gondelach et al., 2014). For these feedstock types, a multitude of conversion technologies exists, ranging from thermo-chemical conversions (e.g., pyrolysis and gasification) over extractions and trans-esterifications to bio-chemical technologies like hydrolysis of lignocellulose and the fermentation of derived sugars (Opia et al., 2021). Since a description of all potential combinations of biomass feedstock and technologies is out of the scope of this thesis, the focus is on only some relevant and industrialized processes.

To substitute conventional products, biomass utilization technologies enable several drop-in points in the already commercialized production chain of plastics. The first drop-in possibility is at the root of plastic production. Pyrolysis and gasification technologies can be used to produce liquid hydrocarbons or synthesis gas from various biomass feedstock types (Pyl et al., 2011; Gollakota et al., 2018; Opia et al., 2021; Holladay et al., 2007). Secondly, biomass can be converted to sugar and subsequently converted to ethanol (Bozell and Petersen, 2010). Already today, ethanol is dehydrated to produce ethylene, one of the major building blocks of current plastics. The commercialized processes are, for instance, operated by Braskem in Brazil to convert sugarcane-based ethanol to ethylene to produce bio-based polyethylene. Third, biomass can be converted to various important platform chemicals via both chemical and biological conversion technologies (Werpy and Petersen, 2004; Bozell and Petersen, 2010). These platform chemicals can subsequently be converted to a multitude of conventional chemical intermediates or directly polymerized to bio-based plastics with a totally different chemical structure than the currently used plastics. Conventional intermediates comprise, for instance, the production of ethylene glycol from bio-based sorbitol (Werpy and Petersen, 2004), or aniline from aminobenzoic acid (Winter et al., 2021). Many intermediates exist for which pilot and commercial plants are currently under development by several companies such as Covestro, UPM, Neste, or Genomatica to name only a few (de Jong et al., 2020). However, which of the several intermediates will make it to industrial scale is yet unclear.

The most well-known example of bio-based plastics produced from bio-based chemicals is polylactic acid (Eş et al., 2018). Polylactic acid is produced from lactic acid obtained by sugar fermentation. Other bio-based plastics are polyhydroxyalkanoates or starch blends (Spierling et al., 2018). Finally, biomass can be used as a source for naturally occurring polymers like cellulose (Muthukumaran et al., 2021) or natural rubber (Sethuraj and Mathew, 1992).

The paragraphs above showed that several circular technologies have already been developed to industrial scale. While reductions of greenhouse gas emissions seem likely, they cannot be taken for granted if these circular technologies are implemented. For instance, converting the inert molecule CO_2 to valuable hydrocarbons, like methanol, requires large amounts of energy, particularly for H_2 production. Thus, overall reductions of greenhouse gas emissions are only warranted if sufficiently low-carbon energy can be provided (von der Assen et al., 2013). In the case of biomass, reduction potentials of greenhouse gas emissions are influenced by the choice of bio-based feedstock and the substituted product such that even an increase of greenhouse gas emissions is possible (Spierling et al., 2018; Muñoz et al., 2014; Cespi et al., 2016). Finally, in the case of plastic recycling, studies indicated that the incineration of high-calorific plastic waste, like polyethylene or polypropylene, instead of coal for energy production leads to a larger reduction of greenhouse gas emissions than mechanical recycling and the substitution of virgin plastic materials (Lazarevic et al., 2010). Thus, to warrant reductions of greenhouse gas emissions by novel circular-carbon technologies, a holistic and standardized method to evaluate environmental impacts should be used: life cycle assessment.

2.3 Life cycle assessment of plastic production and waste treatment

In the previous section, the circular technologies and their aspiration to reduce greenhouse gas emissions of plastics have been described. However, whether these circular technologies reduce greenhouse gas emissions, and actually achieve net-zero greenhouse gas emissions can only be warranted by a sound environmental assessment.

Life cycle assessment is a standardized methodology according to the ISO standards 14040 and 14044 (ISO 14040, 2021; ISO 14044, 2021) for the "compilation and evaluation of the inputs, outputs and environmental impacts of a product system throughout its life cycle". The life cycle describes the production system from the extraction of natural resources, their transportation, over the production of goods and their respective use to the reuse, recycling, or disposal of wastes. Along these life cycle stages, all material and energy exchanges of the production system with the natural environment are quantified. These exchanges with the natural environment are then interpreted regarding their environmental impacts in various categories like climate change (e.g., greenhouse gas emissions), fossil resource depletion, but also human toxicity or eutrophication (Hauschild and Huijbregts, 2015). As a result, life cycle assessment leads to a holistic perspective on the overall environmental impacts and thereby avoids problem shifting between life cycle stages and environmental impacts (Baumann and Tillman, 2004).

The holistic view of life cycle assessment and the avoidance of burden-shifting is essential and fully acknowledged by this thesis. However, the focus of the following literature review on the life cycle assessment of individual circular technologies aims to identify generalized methodological difficulties and gaps. To highlight these methodological difficulties and gaps, greenhouse gas emissions and their impact on global warming are the focus. Thus, benefits and burdens for other environmental impacts are not fully covered by the literature review. Focusing on global warming impacts and greenhouse gas emissions is not altering the conclusions about the general methodological difficulties and gaps.

2.3.1 Fundamentals of life cycle assessment

This section highlights the basic principles of life cycle assessment to provide background information for the following Section 2.3.2 in which life cycle assessments for the circular technologies are presented. The current ISO standards (ISO 14040, 2021; ISO 14044, 2021) include four phases of life cycle assessment (Figure 2.7).

Phase 1: Goal and scope definition
In the first phase, the goal (what?) and scope (what and how?) of the life cycle assessment are defined. Thus, phase 1 is purely question-oriented and does not require any data about physical flows. To define the goal, a clear and unambiguous formulation of the research question must be formulated. To define the scope of the life cycle assessment, the ISO standards require a particularly clear definition of the system boundaries, e.g., what production system is studied, and the so-called functional unit. The functional unit defines the basis to calculate the environmental impacts of a production system, e.g., the production of 1 kg of product A. Additionally, the goal and scope definition defines the target audience, the reason to carry out the study, and whether the results should be disclosed to the public (Baumann and Tillman, 2004).

Phase 2: Life cycle inventory analysis
During the second phase, the complete exchanges with the environment, the so-called

life cycle inventory, are collected. For this purpose, a mathematical model of the production system is developed. This model includes all processes that consume or produce material and energy flows. The material and energy flows are either characterized as elementary or economic flows. Elementary flows either enter or leave the production system without any previous or subsequent human transformation, i.e., these flows are exchanged with the environment. Economic flows occur between human-made processes and, thus, undergo human-made transformation. Furthermore, economic flows are subdivided into products and waste streams. Waste streams have to be treated within the system under study.

Phase 3: Life cycle impact assessment

Based on the life cycle inventory, phase 3 translates and aggregates all elementary flows into environmental impacts categories. Thus, phase 3 is called the life cycle impact assessment. Since the amount of covered elementary flows for one life cycle inventory is often large, several elementary flows have to be aggregated to one environmental impact. For example, the aggregation of the greenhouse gas emission carbon dioxide, methane, or dinitrogen oxide are translated into a single indicator for their global warming impact. For this purpose, characterization factors are used that describe the equivalent effect on a certain environmental impact by different physical substances. To describe the global warming impact of different greenhouse gas emissions, so-called global warming potentials are used. The global warming potential of a single substance measures how much integrated radiative forcing or temperature change is caused during a certain period of time by the emission of 1 kg of the respective substance in relation to 1 kg of CO_2 (IPCC, 2014). Thus, global warming is measured in CO_2-equivalents. The final environmental impact can then be calculated by multiplying the characterization factors for each substance with the amount of each substance consumed or emitted for the production system under study. Note that throughout this thesis, the terms greenhouse gas emissions and global warming impacts are used as synonyms and essentially mean the same, if not stated otherwise. By considering multiple environmental impacts, the problem shifting between several environmental impacts can be avoided (ILCD, 2010a). To select multiple impact categories, it is important to evaluate their respective uncertainty and relevance for the defined goal and scope of the life cycle assessment. A guideline about the selection of environmental impacts is given in the ILCD Handbook by the Joint Research Center of the European Union (ILCD, 2010a).

Phase 4: Interpretation

The fourth and last phase of a life cycle assessment aims to reach conclusions and interpret the results of the study with regard to its limitations. For this purpose, phase 4 starts with the identification of so-called significance issues or hot spots in the respective production system. These hot spots describe the contribution of individual substances or elementary flows to the overall environmental impact (ILCD, 2010a). The second step of phase 4 is related to assessing the robustness and uncertainty of the life cycle assessment results. Uncertainty in life cycle assessment can derive from a multitude of sources, like simplified assumptions and data uncertainty in the life cycle inventory (see phase 2) or characterization factors (see phase 3). To evaluate these uncertainties, sensitivity analysis, Monte Carlo simulations, and analytical methods are used (von Pfingsten et al., 2017).

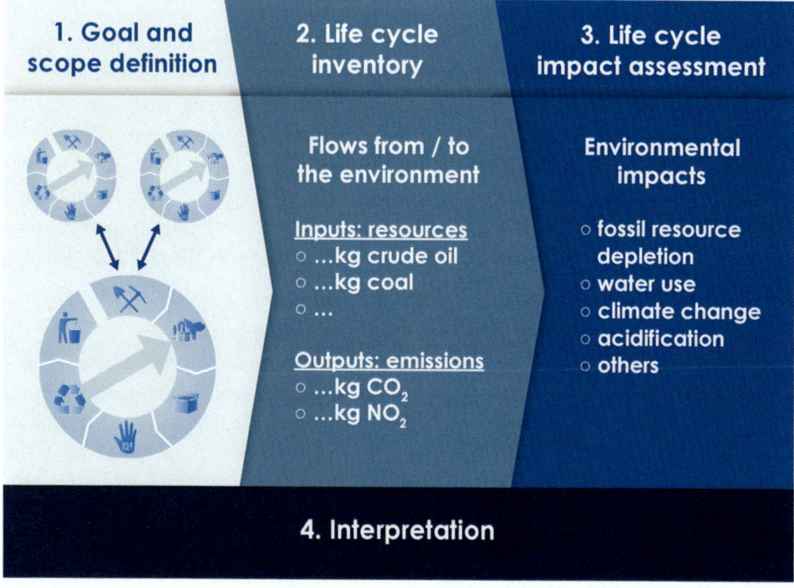

Figure 2.7: The 4 phases of life cycle assessment according to the ISO standards 14040 and 14044 (ISO 14040, 2021; ISO 14044, 2021)

2.3.2 Individual life cycle assessments of circular technologies

The definition of the four phases of life cycle assessment not only plays an important role in the theory and the ISO standards but also in the practical application of life cycle assessment. In practice, life cycle assessment primarily focuses on individual case studies. Already today, a multitude of individual life cycle assessments have been conducted for all three circular technologies, as many review papers highlight for recycling (Antelava et al., 2019; Lazarevic et al., 2010; Bernardo et al., 2016; Blanco et al., 2020; Davidson et al., 2021; Dong et al., 2021; Gomes et al., 2019; Marson et al., 2021; Ramesh and Vinodh, 2020; Xue and Xu, 2017; Zhou et al., 2018; Laurent et al., 2014a), carbon capture and utilization (Artz et al., 2018; Cuéllar-Franca and Azapagic, 2015; Baena-Moreno et al., 2019; Cruz et al., 2021; Garcia-Garcia et al., 2021; Hoppe et al., 2018; Thonemann, 2020; Müller et al., 2020a), as well as biomass utilization (Spierling et al., 2018; Escobar and Laibach, 2021; Hatti-Kaul et al., 2007; Kakadellis and Harris, 2020; Pawelzik et al., 2013; Montazeri et al., 2016).

While the individual case studies provide a very deep understanding of environmental impacts of one particular circular technology, the respective reviews also highlight that methodological choices of the individual life cycle assessment studies hinder the comparability of results and lead to different conclusions (Lazarevic et al., 2010; Laurent et al., 2014b; Montazeri et al., 2016; Spierling et al., 2018; Artz et al., 2018). Due to this variability of results, the following paragraphs will mostly focus on review papers summarizing several studies. By focusing on review articles, a more concise and consistent picture of the current literature about life cycle assessment results for circular technologies can be generated.

In the case of *plastic waste* as renewable feedstock, an early review by Lazarevic et al. (2010) highlighted that the potential to reduce greenhouse gas emissions by mechanical or chemical recycling is mainly dictated by the virgin material substitution ratio and the organic contamination inside the plastic waste. While the virgin material substitution ratio is driven by the quality of the recycled material and dictates the avoided amount of virgin plastic, the organic contamination dictates the demand for sorting and cleaning the waste resource. The higher demand for sorting and cleaning leads to a higher energy demands. The substitution ratio and organic contamination directly impact the greenhouse gas emissions of the recycling technology itself. Furthermore, the potential to reduce greenhouse gas emissions is also influenced by the choice of the benchmark waste treatment and the respective system boundaries (Antelava et al., 2019).
Scientific literature shows that, recycling leads to higher greenhouse gas emissions than incineration with energy recovery if high calorific plastic wastes, like polyethylene, are used to substitute coal in waste incinerators or cement kilns (Patel et al., 2000). However, if natural gas as fossil fuel or the average electricity grid is substituted, incineration of plastic waste was found to lead to higher greenhouse gas emissions than mechanical or chemical recycling (Perugini et al., 2005; Lazarevic et al.,

2010). In contrast, there is no clear picture about the comparison between chemical and mechanical recycling, even though preliminary life cycle assessments indicate a slight advantage of mechanical recycling (Lazarevic et al., 2010).
On a review level, Laurent et al. (2014b) found that many life cycle assessment studies dealing with plastic waste even lack compliance with the ISO standards due to the lack of reporting of used datasets. Additionally, if reported, the used datasets for recycling and other end-of-life options are often outdated or too simplified (Lazarevic et al., 2010). This lack of sufficient documentation and/or the usage of outdated data may, ultimately, even jeopardize the validity of the respective results of life cycle assessments for plastic waste treatment (Laurent et al., 2014a,b).

In contrast to life cycle assessment for plastic waste management that has been assessed since the early 90s (Laurent et al., 2014a), using life cycle assessment for *carbon capture and utilization* is rather a new topic (Thonemann, 2020). However, due to the interest in potential reductions of greenhouse gas emissions, life cycle assessment has rapidly emerged as the tool to evaluate CO_2 utilization technologies from an environmental perspective. Already today, the first guidelines have been released, dealing with the application of life cycle assessment to carbon capture and utilization technologies (Müller et al., 2020b). From a methodological perspective, differences of life cycle assessment studies occur, for instance, due to the functional unit, handling of multi-functionality, or the system boundaries, e.g., if a study covers all or only parts of the product's life cycle. As a result of the extensive literature work, the drivers that dictate the greenhouse gas emissions of carbon capture and utilization technologies are well understood (von der Assen et al., 2013, 2014). The major drivers are highlighted based on the example of methanol, a valuable raw material for polyolefins (Hoppe et al., 2018) or polyurethanes (von der Assen and Bardow, 2014). The upper part of Figure 2.8 visualizes the greenhouse gas emissions per kilogram of methanol with harmonized functional units, handling of multi-functionality, and the system boundaries (Artz et al., 2018).
Even though the methodological choices have been harmonized, the greenhouse gas emissions vary between -1.7 to +9.7 kg of CO_2-eq per kg of methanol. Thus, the respective greenhouse gas emissions not only depend on methodological choices but also largely the respective background datasets for the CO_2 and electricity supply. As a result, the greenhouse gas emissions of CO_2-based methanol can be higher or lower than the fossil-based methanol production based on steam methane reforming.
If the background processes are harmonized, however, the variance in the greenhouse gas emissions of methanol production becomes much smaller. Assuming background datasets representing the electricity and hydrogen supply in 2020, the greenhouse gas emissions per kg of methanol only vary between +1.21 and +1.44 kg CO_2-eq per

kg of methanol. However, the CO_2-based methanol leads to higher greenhouse gas emissions than its fossil-based benchmark. Taking into account wind-based electricity and hydrogen from electrolysis, the range of greenhouse gas emissions for the thermal hydrogenation is even smaller and varies only between -1.28 and -1.27 kg CO_2-eq per kg methanol.
The scientific literature review shows that CO_2-based methanol can only lead to reductions of greenhouse gas emissions if combined with renewable electricity to produce H_2. While methanol is only one example for CO_2-based chemicals, the same trend has been observed for other potential raw materials for plastics, like formic acid, methane, or dimethyl ether (Artz et al., 2018; Thonemann, 2020). One exception is the production of CO_2-based polyols that have been shown to reduce greenhouse gas emissions independently of the carbon intensity of electricity supply (von der Assen and Bardow, 2014).

Considering *biomass* as renewable feedstock, the same problem of comparability of life cycle assessment results has been recognized (Spierling et al., 2018). One particular point of concern in the life cycle assessment of bio-based plastics is the handling of so-called biogenic carbon emissions. Biogenic carbon emissions are those greenhouse gas emissions for which the carbon was once absorbed during plant growth. Some life cycle assessment studies assume that the respective absorbed carbon will be emitted at the end-of-life of the plastic waste and, thus, do not account for the respective carbon uptake during plant growth. As a result, greenhouse gas emissions of producing one kilogram of a bio-based chemical or plastic are often not negative, even though they would be if all greenhouse gas would have been accounted for. This bias in including or excluding biogenic carbon emissions can lead to misinterpretations of the environmental benefits of bio-based chemicals and plastics.
In addition to biogenic carbon, the potential benefits of bio-based chemicals and plastics can largely vary due to the chosen biomass feedstock, the biomass cultivation approach or the allocation criteria for by-products (Pelton, 2019; Renouf et al., 2008; Tsiropoulos et al., 2013). Finally, the chosen benchmark for each bio-based plastic largely influences the respective potential to reduce greenhouse gas emissions. With this regard, results from an individual life cycle assessment study showed that depending on the chosen reference data for polyethylene production, the bio-based counterpart reduces or does not reduce greenhouse gas emissions (Tsiropoulos et al., 2015). Finally, indirect effects on land-use change led to a broad discussion of whether bio-based products actually reduce greenhouse gas emissions or not (Searchinger et al., 2009). Overall, scientific literature shows that if biogenic carbon emissions are included and indirect effects of land-use are avoided, bio-based chemicals and plastics can lead to reductions of greenhouse gas emissions.

The analysis of the review papers revealed that the methodological differences of individual life cycle assessments for all circular technologies include, not exclusively, the choices of functional units, allocation criteria, and system boundaries, but also the use of varying fossil-based benchmarks and datasets for each circular technology. Due to these methodological differences, the individual life cycle assessment studies are often not directly comparable to each other and can lead to contradicting results. In fact, depending on the methodological choices, results of life cycle assessment can show the potential for reductions or increases of greenhouse gas emissions for the same circular technology. Due to these contradicting results, decision-makers often lack clarity whether a particular circular technology is beneficial to reduce greenhouse gas emissions or not. However, the aforementioned literature review also shows that if methodologies and data are aligned, robust conclusions can be drawn about the potential reductions of greenhouse gas emissions by circular technologies (cf. Figure 2.8 for CO_2-based methanol).

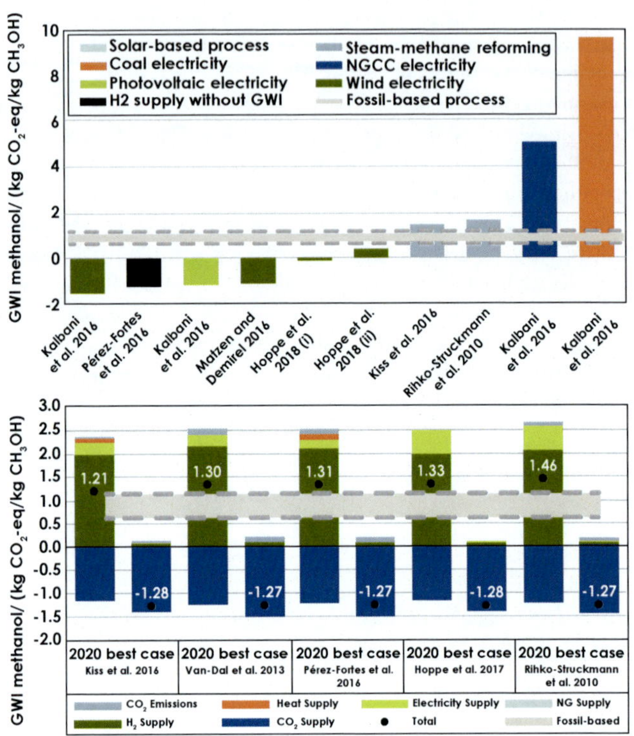

Figure 2.8: The variance of global warming impact (GWI), measured in greenhouse gas emissions, is shown in the upper part for the thermal hydrogenation of CO_2. The lower part of the figure shows the same process concepts but with harmonized methodologies and background datasets for several literature sources that describe the hydrogenation of CO_2 (Al-Kalbani et al., 2016; Matzen and Demirel, 2016; Pérez-Fortes and Tzimas, 2016; Hoppe et al., 2018; Kiss et al., 2016; Van-Dal and Bouallou, 2013). The case "2020 case" assumes that CO_2 is captured from a coal power plant, H_2 is supplied via steam methane reforming, and for natural gas, heat, and electricity the average European values for greenhouse gas emissions are used (Sphera, 2019). The "best-case" refers to an ideal CO_2 source, hydrogen production from electrolysis supplied by wind electricity, natural gas supply as in the "2020-case", as well as heat (95% efficiency) and electricity supplied by wind-based power. Figure according to Artz et al. (2018)

2.3.3 Industry-wide reductions of greenhouse gas emissions

Individual life cycle assessments of plastics provide a very detailed view on particular production technologies and/or products but do not assess the industry-wide implications of using circular technologies to reduce greenhouse gas emissions of plastics. However, this system-wide perspective is important to understand if net-zero emission plastics can be achieved, how such a net-zero future could look like (e.g., which technologies are needed), and what amount of renewable carbon sources would be required to achieve the transition. Thus, in the following paragraphs, industry-wide assessments of plastics or part of its supply chain are summarized.

Currently, the plastic supply chain mainly builds on petrochemicals and, thus, is strongly connected to the oil industry. The oil industry has a long history in generating industry-wide models of its supply chain. For instance, the GREET model first presented by Brown (1996) assesses the well-to-wheel greenhouse gas emissions of passenger cars and trucks. With this regard, the model covers over 100 technologies to produce fuels from fossil- and renewable feedstock. As a result, several models and assessments emerged to assess individual regions, like Europe, and assess measures to reduce greenhouse gas emissions of refinery operations (Berghout et al., 2019; Johansson et al., 2012; Concawe, 2017). For instance, Berghout et al. (2019) found that the combination of carbon capture and storage with biomass gasification could reduce greenhouse gas emissions in north-west Europe by 6.3 Mt CO_2-eq per year, corresponding to a reduction of 154% compared to the base year 2012. Thereby, the combination of biomass utilization and subsequent carbon capture and storage leads to net-negative emissions in the case of refinery operations.

In addition to the regionally focused models, the global greenhouse gas emissions of fuel products have been recently calculated on an unprecedented level of regional and technological detail (Jing et al., 2020). For this purpose, three models have been combined in a consistent manner (Abella and Bergerson, 2012; El-Houjeiri et al., 2013; Masnadi et al., 2018; Young et al., 2019; Gordon et al., 2015): (1) the Oil Production Greenhouse Gas Emissions Estimator (OPGEE), (2) the Petroleum Refinery Life-Cycle Inventory Model (PRELIM) and (3) the Oil Products Emissions Module (OPEM). OPGEE estimates the upstream greenhouse gas emissions from extraction to the refinery input based on detailed data of oil production sites and the respective drilling and manufacturing technologies. PRELIM calculates greenhouse gas emissions of various refinery configurations, depending on the included unit operations, like coking or fluid catalytic cracking, and the crude oil characteristics based on 62 input parameters. OPEM calculates the downstream greenhouse gas emissions from the refinery outlet to the final end use of the oil product, e.g., gasoline, diesel of heavy fuel oils. By combining OPGEE, PRELIM, and OPEM, the authors assess 93% of the global refinery throughput and show that under the projected oil consumption, the industry could save 56-79 Gt CO_2-eq until 2100 by targeting the primary emission sources via carbon capture and storage as well as renewable energy supply (Jing et al., 2020). The very detailed global assessments, however, focus exclusively on petroleum products and only include some chemical feedstock from fluid catalytic crackers (Young et al., 2019). Thus, the models cannot easily be adapted or extended to the industry-wide assessments of chemicals and plastics.

The industry-wide assessment of chemicals and plastics has a long scientific history. Already in the 1970s, early work by Stadtherr and Rudd exploited the possibility to use linear optimization models to assess chemical process networks (Stadtherr and Rudd, 1976), the chemical industry resource demands (Stadtherr and Rudd, 1978) and even the changes of costs and resources demands due to the implementation of new technologies for chemical production (Stadtherr, 1978). Since this early work, various papers have dealt with mathematical aspects of modeling chemical and plastic supply chains (You et al., 2012; Guillén-Gosálbez and Grossmann, 2010). More recently, Thakker and Bakshi (2021) highlighted the need to holistically assess and design entire values chains of products while considering various technological alternatives in a circular network. The authors propose a combination of life cycle assessment and optimization-based approaches to holistically assess the optimal technologies to close life cycle loops.

In contrast to the focus of early work on chemicals and the rather theoretical work about modeling approaches, only some industry-wide case studies assess the potential to reduce greenhouse gas emissions of plastics by circular technologies. Saygin et al. (2014) assessed the global potential of biomass for the production of steam, chemicals, and some polymers in terms of energy savings, CO_2 emissions, cost, and resources availability. For this purpose, the authors use a consistent model of chemical and plastic production to calculate the technical potential of biomass to reduce greenhouse gas emissions. Based on the model, the authors conclude that up to 1.4 Gt CO_2-eq from chemicals and plastics could be abated by the large-scale utilization of biomass. In the same sense, Spierling et al. (2018) analyzed the global potential to reduce greenhouse gas emissions by substituting fossil-based with bio-based plastics using individual life cycle assessment studies. Assuming that two-thirds of the global production volume could be substituted, the authors estimate the potential to be equivalent to 0.32 Gt CO_2-eq. Assessments by Saygin et al. (2014) and Spierling et al. (2018) differ, however, with regard to the scope of chemicals and plastic as well as the underlying individual life cycle assessment studies used for bio-based chemical and plastic production (Section 2.3.2). In addition to the global assessments, Posen et al. (2017) assessed the potential to reduce the greenhouse gas emissions of plastics in the US. The authors highlight that corn-based bio-polymers can reduce industry-wide greenhouse gas emissions by 25%, equal to 16 Mt CO_2-eq per year.

While the aforementioned studies focus on plastic production, the potential contribution to reducing greenhouse gas emissions by improved waste management for plastics has also been assessed in the literature. Due to the local differences in waste management schemes, these analyses focus on regional contributions. For instance, Liu et al. (2018) highlight that, in China, ambitious improvements in recycling rates can reduce 47 Mt of annual greenhouse gas emissions. In the Netherlands potential reductions have been reported to equal 2.3 Mt CO_2-eq annually (Corsten et al., 2013).

Recently, Zheng and Suh (2019) evaluated reductions of greenhouse gas emissions if bio-based and recycled plastics are implemented on a large scale. For this purpose, the authors developed global datasets that cover greenhouse gas emissions of polymer resin production, conversion to plastic products as well as the waste treatment of plastics. The datasets cover over 90% of the global plastic production volume and are based on single life cycle assessment datasets for individual plastics from the database ecoinvent (Ecoinvent, 2020) as well as several individual case studies for bio-based and recycled plastics. Based on their datasets, the authors show that the current way of linear and fossil-based production would increase greenhouse gas emissions from 1.7 to 6.5 Gt CO_2-eq from 2015 to 2050. Considering the combination of recycled and bio-based plastics with renewable energy and a decrease in plastic demands could lead to a reduction of up to 4.8 Gt CO_2-eq per year by 2050. Thus, the authors conclude that greenhouse gas emissions can be kept at today's levels, but the study does not provide a pathway to net-zero emission plastics.

2.4 Scientific gaps for life cycle assessments of plastics

Based on the Sections 2.2 and 2.3, the following major scientific gaps and challenges have been identified.

Gap 1: Making robust decisions in early-development stages
The literature review in Section 2.3.2 shows that the majority of individual product-specific or technology-specific life cycle assessments for circular technologies are based on varying methodological decisions. These methodological choices include functional units, system boundaries, or allocation principles. Methodological choices and the resulting different conclusions lead, in particular, to difficulties for early-stage circular

technologies (Section 2.2.2).
While these difficulties exist for all circular technologies, in particular, chemical recycling to treat plastic packaging waste is intensely debated. In the case of chemical recycling, life cycle assessment results vary strongly and, thus, no clear picture about environmental benefits or burdens has been derived so far. As a result, decision-makers struggle to robustly identify those chemical recycling technologies that will lead to benefits for global greenhouse gas emissions and other environmental impacts. However, identifying the most promising chemical recycling technologies is important before time and money have been invested.
The major challenge lies in the development of a consistent methodology that enables a comparison of multiple chemical recycling technologies while dealing with low data availability in early-development stages.

Gap 2: Lack of technical data and life cycle assessments for plastic products
Due to the complexity of plastic life cycles, a multitude of production technologies exist. For instance, to produce fossil-based chemicals and plastics, over 250 technology routes are have been listed (ICIS, 2018). Considering future technologies based on plastic waste, CO_2, or biomass, the number of technologies is likely to increase. Thus, the data demand for life cycle assessment is generally very high, while the data availability is generally low. Gathering data that describe the complete chemical and plastic supply chain in a sufficient manner for life cycle assessment is an enormous effort. In fact, the generation of appropriate datasets is often the most time-consuming part of a life cycle assessment study (Maranghi and Brondi, 2020).
The generation of technology-specific mass and energy balances that are compliant with the life cycle assessment standard is one of the major challenges identified in the literature. This is particularly the case for complex products and novel supply chains like, for instance, the production of CO_2-based polyurethane products.

Gap 3: Industry-wide and systematic assessment of net-zero emissions plastics
Currently, studies that assess the industry-wide potential to reduce greenhouse gas emissions by circular technologies focus on individual technologies and limited regional scope (Posen et al., 2017; Corsten et al., 2013; Liu et al., 2018; Spierling et al., 2018) or on a global scope but only include part of the available circular technologies (Zheng and Suh, 2019; Saygin et al., 2014). Furthermore, the individual life cycle assessments used to derive an industry-wide picture of plastics often differ in underlying assumptions, data, or methodological choices, and are, thus, not directly comparable (Laurent et al., 2014a; Spierling et al., 2018; Artz et al., 2018).
As a result, studies currently do not systematically evaluate potential technology combinations for net-zero emission plastics and only provide anecdotal evidence of

whether net-zero emission plastics can actually be achieved with the current portfolio of circular technologies (Gabrielli et al., 2020). Individual or partly-combined assessments also neglect potential synergies between the circular technologies as recently reported for polyurethanes (Bachmann et al., 2021). Neglecting these synergies can result in an overestimation of effort to achieve net-zero emission plastics and, thus, hinder the acceptability of respective policy measures due to too high expected cost or energy consumption. An industry-wide and systematic model of all linear and circular technologies is required to solve this challenge, fully depict the pathways to achieve net-zero emission plastics, and provide decision support for all stakeholders.
To build such a model, one particular scientific challenge lies in the gathering of data (cf. gap 2), the harmonizing of these life-cycle-assessment-compliant datasets, and the subsequent combination in a single structured mathematical model. However, a maybe even as complex challenge is using such a detailed model to produce scientifically robust results that are delivered and presented in a clear, structured, and understandable manner.

2.5 Contribution of this thesis

The previous section highlighted the gaps in literature when applying life cycle assessment to circular technologies. The major contribution of this thesis is the development of a consistent, industry-wide, and systematic bottom-up model of the plastic supply chain. However, during the development of this bottom-up model, gaps 1 and 2 have been addressed by two dedicated case studies. In particular, this thesis makes the following contributions:

Robust environmental assessments of chemical recycling
Plastic packaging waste represents over 50% of the global plastic waste volume (Geyer et al., 2017). To increase recycling rates, several novel chemical recycling technologies are advocated to decrease fossil resource depletion and greenhouse gas emissions. To fully assess the environmental benefits of chemical recycling, many different chemical recycling technologies have to be assessed despite low data availability. To do so, this thesis first develops a theoretical model for chemical recycling technologies assuming ideal performance. Subsequently, the thesis generates life cycle datasets for benchmark waste treatment technologies based on industrial data. Based on the assessment of 75 waste treatment scenarios for five environmental impacts, this thesis robustly identifies those chemical recycling technologies that will not result in environmental benefits since already the ideal case has higher environmental impacts than the benchmark waste treatment technology. By this means, Chapter 3 addresses gap 1 by providing

a robust decision-making tool for chemical recycling to scientific literature.

Applying life cycle assessment to CO_2-based rubbers
This thesis applies life cycle assessment to the cradle-to-grave life cycle of CO_2-based rubber products. To do so, novel CO_2-based rubber products are compared against their benchmark products from an environmental perspective. As a result of the analysis, this thesis shows that a system-wide perspective with cradle-to-grave system boundaries is needed to fully evaluate plastic materials and their environmental impacts. Thus, Chapter 4 addresses gap 2 by providing a detailed case study about CO_2-based rubber products to scientific literature.

Global bottom-up model plastic production and waste treatment
Based on the methodological insight during the life cycle assessment of chemical recycling and CO_2-based rubbers, the first-ever, industry-wide, and global bottom-up model of plastic production and waste treatment is generated. This bottom-up model is based on an extensive data generation and harmonization effort that was needed to provide sufficiently detailed technical data. The included data represents over 90% of the global plastic products. Thus, the bottom-up model enables the assessment of the complete supply chain from chemical feedstock, e.g., naphtha over chemical and plastic production to the plastic waste treatment in a consistent and industry-wide manner. As a result, Chapter 5 addresses gaps 1 to 3 by including the major early-stage and mature circular technologies inside the bottom-up model as well as covering over 90% of the global production and waste treatment chain of plastics

Systematic evaluation of pathways to achieve net-zero emission plastics
The bottom-up model is additionally used to calculate the global supply chain, energy demands, and greenhouse gas emissions for the fossil-based and linear plastic life cycle in 2015 and 2050 (Chapter 6). The result for 2015 is benchmarked with available literature sources about the current state of the art plastic production.
In the final Chapter 7, the bottom-up model is used to evaluate how circular technology can achieve net-zero emission plastics. The results show that net-zero emission plastics can only be achieved by combing biomass and CO_2 utilization with an effective recycling rate of over 70%. By this means, Chapters 6 and 7 address gap 3 by showing that net-zero emission plastics can be achieved with 53% lower energy demands and up to 288 billion USD of operational cost savings under favorable market conditions. Thus, this thesis breaks with the paradigm that circular and net-zero emission plastics can only be achieved by increasing energy and costs.

Chapter 3

The environmental potential of chemical recycling for packaging wastes

Currently, almost half of the globally produced plastic waste consists of 5 types of plastic packaging: (1) polyethylene terephthalate (PET), (2) lowdensity polyethylene (LDPE), (3) high-density polyethylene (HDPE), (4) polypropylene (PP) and (5) polystyrene (PS). In 2015, these types of plastic packaging wastes amounted to 141 Mt and the amounts of plastics are expected to further increase (Geyer et al., 2017). Thus, proper handling and recycling of plastic packaging waste is a key challenge to evolve into an environmental-friendly future (Foundation, 2016).

To efficiently recycle large amounts of plastic packaging wastes, a circular economy requires suitable technologies, such as chemical recycling (Rahimi and García, 2017). Chemical recycling turns plastic packaging waste into chemical products, avoiding their production from fossil feedstock in the first place. Therefore, chemical recycling is expected to decrease the demand for the planets' finite fossil resources as well as the emissions of greenhouse gases (Foundation, 2016). However, the environmental benefits of chemical recycling are intensely debated: Geyer et al. (2016) show that closed-loop recycling systems have no intrinsic environmental benefit over open-loop recycling systems. In fact, Shen et al. (2010) found that linear recycling pathways for PET by mechanical recycling are environmentally superior to circular pathways by chemical recycling to feedstock monomers, even if the mechanically recycled PET is ultimately incinerated.

Thus, there is a mismatch between expected environmental benefits and results from prospective environmental evaluations for chemical recycling. The most established method for environmental assessments of products and technologies is a standardized life cycle assessment (ISO 14040, 2021; ISO 14044, 2021). Reviews of life cycle assessment studies for plastic waste treatment showed that methodological variations hinder comparisons of individual case studies and that most data about end-of-life treatment of plastic waste is outdated or not fully reported (Laurent et al., 2014a; Lazarevic et al., 2010) (cf. Section 2.3.2). Thus, methods to assess the environmental benefits of chemical recycling have to overcome two challenges: (1) the assessment of many end-of-life options despite low data availability and (2) a consistent life cycle assessment methodology to ensure comparability.

In this thesis, both challenges are addressed by determining the maximum environmental benefits of 26 chemical recycling technologies compared to 18 benchmark waste treatment technologies based on a consistent life cycle assessment methodology. These maximum benefits are denoted the "environmental potential of chemical recycling". For this purpose, a life cycle assessment method is derived that is based on reaction chemistry and ideal thermodynamic data for chemical recycling. The life cycle assessment method is used to assess the environmental potential of chemical recycling for the major plastic packaging wastes: PET, HDPE, LDPE, PP, and PS. The environmental impacts of (1) ideal chemical recycling are compared to (2) the benchmark waste treatment. The benchmark waste treatment is based on industrial and up-to-date datasets for mechanical recycling and waste incineration with energy recovery. This ideal thermodynamic assessment leads to minimal environmental impacts for chemical recycling. Thus, chemical recycling without any promise of environmental benefits is identified since their minimal environmental impacts are already higher than those of current benchmark waste treatments.

In Section 3.1, the materials and methods to calculate the environmental potential of chemical recycling are presented. Section 3.2 presents and discusses the results of the study. Conclusions about the environmental potential of chemical recycling technologies are presented in Section 3.3.

Major parts of this Chapter are reproduced with permission of Elsevier from:

Meys, R., Frick, F., Westhues, S., Sternberg, A., Klankermayer, J., and Bardow, A. (2020). Towards a circular economy for plastic packaging wastes - the environmental potential of chemical recycling. *Resources, Conservation and Recycling*, 162:105010.

The author of this thesis contributed to the development of the concept of the study (including the Supporting Information, cf. Appendix A), performed the theoretical modeling, was key responsible for data curation and result interpretation, wrote the original draft, and edited the final manuscript.

3.1 Materials and methods

The environmental potential is based on a comparative life cycle assessment between chemical recycling and its benchmark waste treatment. In this Section, the general methodology to calculate the environmental potential is presented: the system boundaries and the functional unit, the calculation of ideal chemical recycling inventories, the relevant impact categories, the uncertainty and robustness of results, and the scope of the analysis. Values for all inventories are given in Appendix A.

3.1.1 System boundaries and functional unit

Life cycle assessment evaluates the environmental impacts of products and processes from cradle-to-grave, e.g., from the extraction of raw materials over the use-phase to the disposal or recycling (Baumann and Tillman, 2004). In comparative life cycle assessments, only changing activities need to be considered because identical activities cancel each other out in all comparisons (Eriksson et al., 2005). For instance, waste generation can be neglected when comparing different energy recovery options for the same plastic waste. Here, it is assumed that waste is collected and sorted before entering waste treatment. Thus, collection and sorting can be omitted in the comparative life cycle assessment study.

Chemical recycling has two major effects: the benchmark waste treatment (e.g., landfill) is substituted and primary chemical production is avoided. However, the benchmark waste treatment might produce a valuable product, which now needs to be compensated by primary materials. Thus, system boundaries need to include the environmental impacts of (1) the benchmark waste treatment (WT) or chemical recycling (CR) and (2) the avoided benchmark product (avP) or the avoided chemical production (avC) (cf. Figure 3.1). The avoided production leads to a credit in life cycle assessment, denoted as "avoided burden", and is commonly used in life cycle assessments for waste treatment.

In life cycle assessment, the functional unit ensures a consistent comparison between technology options (Baumann and Tillman, 2004). Previous life cycle assessment studies highlighted that the composition of plastic waste affects life cycle assessment results (Lazarevic et al., 2010). Thus, the functional unit is defined as the treatment of 1 kg of plastic packaging waste of defined composition, including organic/inorganic contamination and moisture, entering the waste treatment technology.

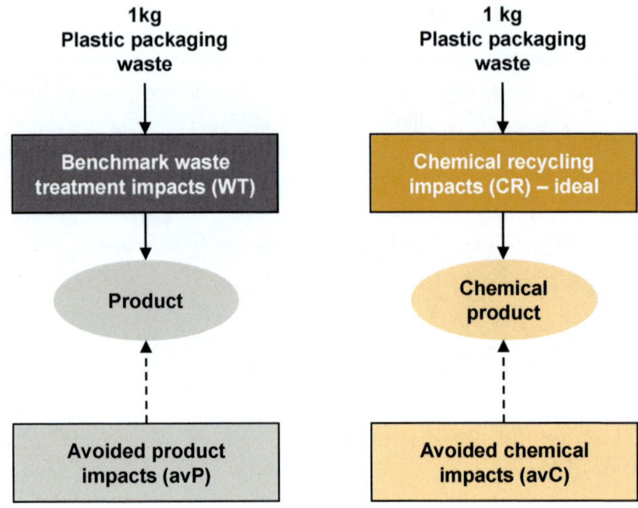

Figure 3.1: Comparison between the benchmark waste treatment (WT) and chemical recycling (CR) including the avoided conventional products (avP) and chemicals (avC). The input stream is 1 kg collected and sorted plastic packaging wastes.

3.1.2 Computation of the environmental potential

The environmental potential (E_{pot}) is the difference between the net environmental impacts of the benchmark waste treatment ($EI_{WT,net}$) and ideal chemical recycling ($EI_{CR,ideal,net}$), equation (3.1). Net environmental impacts consist of direct environmental impacts of the benchmark waste treatment ($EI_{WT,direct}$) and chemical recycling ($EI_{CR,ideal,direct}$) minus the credit for avoided products (EI_{avP}) and chemicals (EI_{avC}):

$$E_{pot} = \overbrace{(EI_{WT,direct} - EI_{avP})}^{EI_{WT,net}} - \overbrace{(EI_{CR,ideal,direct} - EI_{avC})}^{EI_{CR,ideal,net}} \tag{3.1}$$

The direct environmental impact of the benchmark ($EI_{WT,direct}$) includes all environmental impacts required to treat 1 kg of plastic packaging waste. For example, energy recovery from plastic packaging waste leads to direct environmental impacts from flue gas emissions, the supply of auxiliary materials and energy for flue gas cleaning, as well as residue treatment (fly ash, sludge, waste water, and slag treatment)

(Doka, 2003, 2013). Avoided environmental impacts (EI_{avP}) of energy recovery include a credit for the products heat and electricity (Eriksson and Finnveden, 2017). For all benchmarks, inventories are taken from industrial datasets (cf. Section 3.1.6).

Chemical recycling converts 1 kg of treated plastic packaging waste into m_j kg of one or more chemicals. These chemicals substitute their conventional production and the corresponding environmental impacts (EI_j), leading to a credit (EI_{avC}):

$$EI_{avC} = \sum_{j=1}^{n} EI_j m_j \tag{3.2}$$

Environmental impacts (EI_j) are based on data from life cycle assessment databases and represent the production of identical chemicals from cradle-to-gate. For chemically identical products, the gate-to-grave phase is the same and does not lead to further credits if market-mediated effects can be excluded.

The remaining challenge is to calculate the direct environmental impacts of the ideal chemical recycling ($EI_{CR,ideal,direct}$), because only limited data is available, e.g., only the reaction equations. However, this limited data is sufficient for the analysis since the goal is to compute the minimal environmental impacts of chemical recycling. For this purpose, an approach based on stoichiometry and thermodynamic data is proposed. The approach includes (1) the reactants ($\sum_{i=1}^{o} EI_i m_i$), (2) residual wastes ($\sum_{k=1}^{p} EI_k m_k$), and (3) thermal energy ($EI_H Q_H$):

$$EI_{CR,ideal,direct} = \sum_{i=1}^{o} EI_i m_i + \sum_{1}^{p} EI_k m_k + EI_H Q_H \tag{3.3}$$

Here, m is the respective mass of reactants i and residual wastes k. $EI_{i/k}$ represents the corresponding environmental impacts per kg to produce reactant i or treat residual waste k. Q_H is the thermal energy required for chemical recycling per 1 kg of treated plastic packaging waste. EI_H denotes the environmental impact of providing 1 unit (e.g., per MJ or kWh) of thermal energy.

The goal is to calculate the minimal environmental impacts of chemical recycling. In line with this goal, Q_H is the minimal energy demand for the complete chemical recycling process per 1 kg of plastic packaging waste. It is assumed that energy is provided by heat because the energy demand is largely driven by the chemical decomposition of polymers that takes place above 300°C (Lopez et al., 2017; Rahimi and García, 2017). Using electric heating above these temperatures is possible, however, it would lead to higher global warming impacts under current emission intensities of electricity supply. Thus, EI_H denotes the environmental impact of providing 1 MJ of thermal energy by industrial furnaces using natural gas as fuel since natural gas is the most commonly used fuel for industrial processes in Europe (Ecoinvent, 2020). All masses and the required thermal energy are calculated assuming ideal thermodynamics:

(I) 100 % conversion of polymer waste based on the stoichiometric reaction

$$\underbrace{\sum_{1}^{i} v_i M_i}_{reactants} + \underbrace{v_p M_p}_{polymer} = \underbrace{\sum_{1}^{j} v_j M_j}_{products} \tag{3.4}$$

(II) Minimal energy demand Q_H based on the reaction enthalpy ΔH_R^0 at standard conditions (25 °C, 1 atm):

$$Q_H = \Delta H_R^0 \tag{3.5}$$

where the reaction enthalpy H_R^0 at standard conditions (25°C, 1 atm) is the difference of the standard enthalpies of formation of the pure in- and outputs (see Table A.6 in Appendix A.9).

3.1.3 Environmental impacts

In this chapter, the focus is on global warming and fossil resource depletion because these are stated as the major targets of a circular economy for plastic waste (Foundation, 2016). Additionally, terrestrial acidification and marine/freshwater eutrophication are studied (cf. Appendix A, Sections A.15 and A.16). All impact categories are modeled according to Recipe midpoint indicators implemented in Ecoinvent (Ecoinvent, 2020). It is important to keep in mind that the physical lower bound of the environmental impacts depends on the method and version used for impact assessment. Furthermore, note that the global warming impact category is measured in CO_2-equivalent emissions (CO_2-eq), which equals the total amount of greenhouse gas emissions. As a result, if this thesis refers to global warming impacts, it is equal with the accumulated sum of all greenhouse gas emissions (cf. Section 2.3).

3.1.4 Robustness and uncertainty of the environmental potential

Full life cycle assessments require, at best, industry-based inventories. Based on these inventories, the environmental benefits (EI_{ben}) by chemical recycling could be calculated as the difference between the benchmark ($EI_{WT,net}$) and the industrialized chemical recycling ($EI_{CR,net}$):

$$EI_{ben} = EI_{WT,net} - EI_{CR,net} \tag{3.6}$$

However, many chemical recycling technologies are in early development and sufficient data is missing. To still analyze chemical recycling, a thermodynamically ideal chemical recycling process is assumed (cf. Section 3.1.2). The ideal assumptions lead to an optimal chemical recycling process with minimal environmental impacts, $EI_{CR,ideal,net}$. Real-case chemical recycling $EI_{CT,net}$ will always lead to increased environmental impacts such that $EI_{CT,net} > EI_{CR,ideal,net}$. For instance, ideal thermodynamic models result in minimal energy demands. Real chemical recycling would increase energy demands and environmental impacts.

As a result, the environmental potential ($EI_{pot} = EI_{WT,net} - EI_{CR,ideal,net}$, cf. Section 3.1.2, eq. 3.1) will always be higher than the environmental benefits of chemical recycling, i.e., the following equation is always true for the same type of energy supply: $EI_{pot} > EI_{ben}$.

Thereby, the analysis allows identifying the most promising chemical recycling pathways. However, a positive environmental potential does not imply that a chemical recycling pathway is beneficial but that it can be better than the benchmark. A full life cycle assessment is warranted to ensure the benefits of the real chemical recycling technology. The uncertainty of the obtained results is assessed in a sensitivity analysis (cf. Section 3.2.3). This sensitivity analysis varies the energy demand and the conversion rate of chemical recycling.

In contrast, the method is robust for negative environmental potentials: a negative value indicates that chemical recycling is inferior to the benchmark even for ideal conditions. Thus, for real conditions, no benefits can be expected because environmental impacts can only increase since $EI_{pot} > EI_{ben}$ is always true.

3.1.5 Scope of the study

Packaging represents approx. 50% (Geyer et al., 2017) of global plastic waste. The study covers five major plastic packaging wastes: PET, HDPE, LDPE, PP, and PS. The literature review identified 26 chemical recycling technologies and 18 benchmark waste treatments, e.g., waste incineration, energy recovery in cement kilns, and mechanical recycling for each plastic packaging waste. All packaging wastes are assumed to be collected and sorted before the waste treatment. Furthermore, the environmental potential is calculated in the European context, where the circular economy is gaining increased attention (European Commission, 2015) and sufficient data is available. In this chapter, the technologies are only briefly described. Details can be found in Appendix A.

3.1.5.1 Benchmark waste treatments

Waste treatment of plastic packaging waste is dominated by three technologies: landfill, energy recovery, and mechanical recycling (Geyer et al., 2017). In 2015, landfilling represented the dominating technology and accounted for 59% of the treatment of plastic waste (Geyer et al., 2017). However, landfill bans are discussed in several world regions and were proven highly effective. In Europe, landfill rates dropped by 38% and recycling rates increased by 64% from 2006 to 2014. Based on the success, several European countries are planning to implement landfill bans (PlasticsEurope, 2018). Thus, landfilling is expected to play a minor role in Europe and, thus, is not considered as a benchmark waste treatment.

If landfilling is banned, energy recovery is currently the only option for plastic packaging waste that cannot be recycled mechanically. In principle, energy recovery can be performed by (1) incineration in municipal solid waste incinerators or (2) co-combustion with other fuels (Lazarevic et al., 2010). When assessing the environmental impacts of energy recovery, the energy production that is substituted has to be carefully assessed (Eriksson et al., 2007). In regions with cement industries, plastic packaging waste is often co-combusted in cement kilns to reduce lignite or hard coal utilization (Dehoust and Christiani, 2012). However, in countries where no cement industry exists, plastic packaging waste is incinerated in municipal solid waste incinerators for combined heat and power production (Eriksson and Finnveden, 2009). Here, both scenarios for energy recovery are included.

The first scenario substitutes coal, e.g., lignite, as fuel in cement kilns. This assumption is based on the large share of coal in cement plants' fuel supply of 70% in 2017 (IEA and CSI, 2018). For the future, even the most sustainable scenario of the International Energy Agency (IEA) predicts that in 2030, 55% of the fuel used in the cement industry is coal. Despite the important role of coal as fuel in the cement industry, alternative fuels are expected to increase: For 2030, the IEA predicts a share of 18% of biomass, waste, and other renewable fuels as well as 12% of natural gas in the fuel mix of cement plants. To account for future fuels shifts in the cement industry, two alternative scenarios are considered in which either natural gas or biomass, e.g., waste wood pellets, are substituted as fuel in cement plants. In the second scenario, plastic packaging waste is incinerated in a municipal solid waste incinerator with an energy efficiency of 41% and a power to heat ratio of 0.35. These values represent the average efficiencies for municipal solid waste incinerators in Europe (Eriksson and

Finnveden, 2009). The products heat and electricity are assumed to substitute average European grid electricity and district heating (Ecoinvent, 2020).
Mechanical recycling currently plays a minor role in global waste treatment: today, only 16% of plastic packing waste is recycled mechanically (Geyer et al., 2017). In contrast, a recent survey for Germany in 2015 showed that up to 73% of collected and sorted plastic packaging waste was recycled mechanically (Consultic, 2015). Thus, if plastic packaging waste is collected and sorted, it can be efficiently recycled. However, the recycled granulates cannot fully substitute virgin polymer granulates due to losses of product properties. To account for this downcycling, a so-called substitution factor is used frequently (BIO Intelligence Service, 2013; Michaud et al., 2010; Prognos AG, 2008). The substitution factor represents the amount of virgin plastic that can be substituted by 1 kg recycled plastic (e.g., technical quality measures). The substitution factor typically ranges from 0.7 for HDPE, LDPE, and PP to 1 for PET (BIO Intelligence Service, 2013; Michaud et al., 2010; Prognos AG, 2008).

3.1.5.2 Chemical recycling

Chemical recycling was recently discussed by several excellent reviews (Rahimi and García, 2017; Clark et al., 2016; Hong and Chen, 2017; Ragaert et al., 2017; Anuar Sharuddin et al., 2016; Singh et al., 2017). Based on the provided information, 4 categories for chemical recycling are derived: (1) use as refinery feedstock, (2) fuel production, (3) monomer production, and (4) chemical upcycling. Note that LDPE, HDPE, and PP are collectively denoted polyolefins in the following due to their similar properties in chemical recycling. The four categories of chemical recycling are described in the following and summarized in Table 3.1:

(1) **Refinery feedstock:** Plastic packaging wastes can be liquefied and substitute crude petroleum oil and intermediates, such as naphtha, in refinery and steam cracking processes. More precisely, polyolefins and PS can be utilized (Lopez et al., 2017), whereas PET results in large amounts of corrosive benzoic acid (up to 0.5 kg per kg PET). Benzoic acid can block pipes and heat exchangers (Hong and Chen, 2017). Hence, PET as a refinery feedstock is excluded. A substitution on an equivalent mass basis is assumed. This assumption is justified by the small difference of carbon and hydrogen contents[1] of polyolefins (C: 85.63 wt.%, H: 14.37 wt.%) and polystyrene (C: 92.26 wt.%, H: 7.74 wt.%) compared to petroleum oil (C: 85 wt.%, H: 12 wt.%) (Ecoinvent, 2020). Due to the small difference, no substantial change of products from refinery units is expected by

[1]Calculated based on chemical composition of raw materials ethylene, propylene, and styrene.

utilizing either polyolefins, PS, or petroleum oil/naphtha.

(2) **Fuel production:** Even though fuel production might not be considered "full recycling", it is considered in the subsequent analysis because at least the plastics energy content is re-used, the production of fossil fuels is avoided, and many reviews of chemical recycling include fuel production. Fuel production decomposes the molecular structure of the polymer by gasification or pyrolysis and results in mixtures of gaseous fuels ($C_1 - C_{5/6}$) (Honus et al., 2018a,b) or liquid fuels ($C_5 - C_{20}$) (Lopez et al., 2017). For gaseous fuels, yields of up to 99% (Li et al., 1999) are reported for polyolefins, while PET and PS result in lower yields of 76.9% and 9.9%, respectively (Anuar Sharuddin et al., 2016). The low yields for PS are due to the aromatic fraction in PS. This aromatic fraction produces solids and liquids during gasification and, thus, decreases gaseous fuel yields (Honus et al., 2018a,b). For PET-derived gaseous fuels, gasification leads to high CO_2-concentrations, which reduce net calorific values to only up to 9.7 MJ per kg (Honus et al., 2018a,b) compared to natural gas with 50.4 MJ per kg (Ecoinvent, 2020). Thus, PET-derived gaseous fuels are excluded from the study. The considered gaseous fuels are expected to substitute natural gas based on an equal net calorific value. Liquid fuels can be produced from polyolefins with yields of up to 95.7% (Zeaiter, 2014). For PET and PS, yields of 38.89% and 96.73% have been reported (Anuar Sharuddin et al., 2016). Liquid fuels are expected to substitute diesel and gasoline based on the net calorific value (Kalargaris et al., 2017).

(3) **Monomer production:**

Polyolefins can be recycled back to monomers by thermal or catalytic pyrolysis. High yields of up to 75% (Milne et al., 1999) can be achieved for conversion to ethylene and propylene by utilizing high temperatures above 800°C and short residence times or highly selective catalysts such as HZSM-5, HY, and Hβ zeolites at 500°C (Elordi et al., 2009). For PET, many pathways have been proposed to produce feedstock monomers: e.g., hydrolysis, alcoholysis, acidolysis, or aminolysis (Carta et al., 2003; Paszun and Spychaj, 1997). Most pathways apply transesterification reactions with nucleophiles such as ethylene glycol or methanol to produce terephthalic acid or dimethyl terephthalate together with ethylene glycol (Rahimi and García, 2017). Conversion rates of 90% were reported for the production of both terephthalic acid and dimethyl terephthalate (Paszun and Spychaj, 1997). For PS, the most efficient chemical recycling technologies to monomers employ metal oxides such as MgO, BaO, CaO, or K2O. The resulting yield to styrene is approx. 70% (Zhang et al., 1995). The production of monomers is assumed to substitute an equal mass of monomers because chemically identical molecules are produced.

(4) **Chemical upcycling:** Plastic packaging waste can be converted to value-added chemicals by so-called chemical upcycling. Recent advances show that molecular ruthenium catalysts enable selective chemical upcycling of mixed plastic wastes (e.g., PET/PLA) to value-added chemicals, such as 1,4-benzenedimethanol and ethylene glycol (Westhues et al., 2018). Furthermore, aminolysis or glycolysis of PET can produce bis(2 hydroxyethyl) terephthalate (Paszun and Spychaj, 1997). Bis(2-hydroxyethyl) terephthalate and 1,4-benzenedimethanol are regarded as promising monomers for cyclohexane di-methanol (Westhues et al., 2018; Guo et al., 2015). Cyclohexane di-methanol is used in the polymer industry for value-added polyester fibers (Turner, 2004) and as a polycarbonate substitute (Ritter, 2011). For polyolefins and PS, no chemical upcycling technologies have been so far proposed to the best of the authors' knowledge.

Table 3.1: Overview of substituted chemicals produced by chemical recycling of plastic packaging waste.

	category	PET	PS	polyolefins (LD/HDPE+PP)
(1)	refinery production	excluded from study	crude petroleum oil	crude petroleum oil
(2)	fuel production	gasoline, and diesel	natural gas, gasoline, and diesel	natural gas, gasoline and diesel
(3)	monomer production	terephthalic acid or dimethanol terephthalate; both with ethylene glycol	styrene	ethylene or propylene
(4)	chemical upcycling	cyclohexane dimethanol and ethylene glycol	not available	not available

3.1.6 Data requirements and used datasets

All details about datasets can be found in Appendix A. Waste compositions were obtained from industrial sorting facilities (DSD, 2018). For mechanical recycling, environmental impacts are based on current industry datasets from recycling practitioners from the late-year 2017 (BIO Intelligence Service, 2013; HTP GmbH & Co. KG, 2017; Prognos AG, 2008). For energy recovery, the environmental impacts is calculated according to (Doka, 2003, 2013). Modeling principles and assumptions, as well as thermodynamic properties required to calculate the environmental impacts for chemical recycling, are summarized in Sections A.1 to A.10 of Appendix A. All background datasets are obtained from life cycle assessment databases for supplied materials (e.g., reactants) and energy carriers (e.g., heat and electricity) and avoided conventional products, except for lignite, diesel, and gasoline (Ecoinvent, 2020; Sphera, 2019).

In life cycle assessment, the quality of data used to model the technical processes is of utmost importance since it dictates the quality and robustness of the results. Data about mechanical recycling technologies available in the literature is frequently outdated and based on simplified assessments (Laurent et al., 2014a,b). In contrast, the data about mechanical recycling technologies used in this thesis are based on industry datasets and more up-to-date years. The data for waste incineration and energy recovery in cement kilns is based on well-established modeling principles that also include industry data from Germany and Switzerland (Doka, 2003, 2013). Additionally, all datasets for substituted products were taken from standard and well-accepted life cycle assessment databases. Thus, for the benchmark waste treatment the overall data quality can be assumed high. For a data quality and uncertainty evaluation of chemical recycling see Section 3.1.4.

3.2 Results and discussion

3.2.1 Global warming impacts and fossil resource depletion

Results are discussed for each benchmark waste treatment separately. For fossil resource depletion, the environmental potential is almost always positive. Thus, most chemical recycling technologies have the potential to reduce the use of fossil resources. Therefore, the following discussion focuses on global warming impacts if not mentioned otherwise.

Chemical recycling vs. energy recovery in municipal solid waste incinerators

All chemical recycling technologies have a positive environmental potential compared to energy recovery in waste incinerators (see Figure 3.2). The environmental potential for global warming ranges from 0.78 kg CO_2-eq for gaseous fuels from PS to 4.21 kg CO_2-eq for chemical upcycling of PET to cyclohexane di-methanol. Chemical recycling could potentially reduce global warming because municipal solid waste incinerators receive low credits for electricity and heat production. Credits for global warming impacts are low since large amounts of electricity in Europe are already produced from renewables. Furthermore, district heating is mainly based on natural gas, which represents the least harmful fossil source of heat (Connolly et al., 2014). In the near-term future, electricity production and district heating will be further decarbonized and, thus, the environmental impacts avoided by municipal solid waste incinerators will further decrease. This trend will further increase the environmental potential for all chemical recycling technologies compared to municipal waste incinerators.

Chemical recycling vs. energy recovery in cement kilns

Environmental potentials for global warming are less homogeneous for the comparison with energy recovery in cement kilns (cf. Figure 3.3). Note that, in cement kilns, plastic packaging waste is assumed to substitute lignite if not stated otherwise. The results differ for two groups of chemical recycling: (1) refinery feedstock and fuel production and (2) monomer production and chemical upcycling.

Recycling plastic packaging waste to refinery feedstock and fuels has negative environmental potentials ranging from -1.46 kg CO_2-eq for gaseous fuels from HDPE to -0.44 kg CO_2-eq for liquid fuel production from PET. Negative environmental potentials result from two effects: (1) using plastic packaging waste as fuel in cement kiln results in high credits from substituting lignite and (2) chemical recycling of plastic packaging waste to refinery feedstock or fuels results in low credits from substituting fossil products. Thus, the substitution of lignite is favorable over the production of refinery feedstock or liquid and gaseous fuels.

Monomer production results in negative environmental potentials for HDPE and LDPE, a small positive potential for PP and positive environmental potentials for PET and PS. The results are based on two opposite effects: (1) the higher net calorific values and, thus, higher credit for lignite substitution of HDPE, LDPE, PP and PS compared to PET and (2) the lower credits for monomer production from HDPE, LDPE and PP compared to PET and PS.

Chemical recycling of HDPE and LDPE achieves a low credit for the global warming impacts avoided from the conventional production of ethylene. At the same time, high calorific HDPE and LDPE must be compensated by lignite in cement kilns resulting in higher emissions. As a result, environmental potentials are negative. For PP, the increase in avoided global warming impacts in cement kilns is slightly less than for

negative potential	positive potential	Global warming impact [kg CO_2-eq]					
		refinery feedstock	gaseous fuels	gasoline	diesel	monomers	chemical upcycling
(1) energy recovery in waste incinerator	PET TPA			0.90	0.92	3.01	4.21
	PET DMT					3.67	
	LDPE	1.34	0.88	1.67	1.72	2.11	
	HDPE	1.38	0.91	1.71	1.77	2.17	
	PP	1.45	0.93	1.72	1.77	2.30	
	PS	1.70	0.78	1.51	1.55	3.66	

negative potential	positive potential	Fossil resource depletion [kg oil eq]					
		refinery feedstock	gaseous fuels	gasoline	diesel	monomers	chemical upcycling
(1) energy recovery in waste incinerator	PET TPA			0.32	0.36	1.10	1.55
	PET DMT					1.34	
	LDPE	0.51	0.46	0.61	0.68	0.83	
	HDPE	0.52	0.47	0.62	0.69	0.84	
	PP	0.54	0.47	0.62	0.69	0.91	
	PS	0.58	0.41	0.55	0.61	1.34	

Figure 3.2: Environmental potential for global warming impacts and fossil resource depletion of chemical recycling compared to energy recovery in municipal solid waste incinerators. Red indicates negative environmental potentials. Green indicates positive environmental potentials. White indicates values closer to zero. Grey indicates that chemical recycling does not exist or has been omitted. PET can be used to produce ethylene glycol and two types of monomers: terephthalic acid (PET TPA) and dimethyl terephthalate (PET DMT).

HDPE and LDPE. Thus, chemical recycling to propylene has a small positive environmental potential. For PS, the global warming impacts avoided from conventional styrene production are higher than for HDPE, LDPE, and PP and, thus, outweigh the increase of cement kiln impacts due to increased lignite utilization. Thus, monomer production from PS has a positive environmental potential.

For PET, low net calorific values lead to small avoided global warming impacts from lignite utilization. At the same time, credits are high for replacing conventional terephthalic acid/dimethyl terephthalate production. Thus, the environmental potential of monomer production from PET is positive. Furthermore, the environmental potential for PET reaches up to 2.88 kg CO_2-eq if cyclohexane di-methanol is produced due to the very high avoided global warming impacts of conventional cyclohexane di-methanol production.

From the perspective of global warming, energy recovery in cement kilns is superior to chemical recycling to refinery feedstock and fuels for all polymer types if lignite can be substituted in cement kilns. Thus, sorted plastic packaging waste should rather be used as a substitute for lignite in cement plants than chemically recycled to refinery feedstock or fuel products. For LDPE and HDPE, energy recovery in cement kilns is even preferable to ideal chemical recycling to monomers. In contrast, PET and PS, otherwise being used in cement kilns, seem promising candidates for monomer production and chemical upcycling.
In contrast to lignite usage, chemical recycling seems more promising if biomass or natural gas would be replacing plastic packaging as fuel in cement kilns (cf. Figure S7 in the Supplementary Material). The only chemical recycling technology still resulting in negative environmental potentials for both global warming impacts and fossil resource depletion is recycling gaseous fuels due to the low credit for substituting natural gas.
For all other chemical recycling technologies, except the production of gaseous fuels, the environmental potentials range from 0.35 kg CO_2-eq for diesel from PET to 3.77 kg CO_2-eq for chemical upcycling of PET to cyclohexane di-methanol if plastic packaging waste replaces natural gas as fuel. If plastic packaging waste would replace biomass, positive environmental potentials reach up to 5.06 kg CO_2-eq for chemical upcycling of PET to cyclohexane di-methanol. The positive environmental potentials are based on the fact that natural gas and biomass are less emission-intensive fuels than plastic packaging waste in cement kilns. Thus, plastic packaging waste should be rather for chemical recycling than to substitute natural gas or biomass in cement kilns.

3.2.1.1 Chemical recycling vs. mechanical recycling

The comparison between chemical recycling and mechanical recycling (cf Figure 3.4) can again be subdivided into two groups: (1) refinery feedstock and fuel production with negative environmental potentials and (2) monomer production and chemical upcycling with positive environmental potentials. The results are mainly based on high global warming impacts avoided by mechanical recycling. To compete with mechanical recycling, chemical recycling must avoid high global warming impacts from conventional chemical production.
Chemical recycling to refinery feedstock and gaseous fuels avoids only small global warming impacts to produce crude petroleum oil naphtha and natural gas, respectively. For chemical recycling to liquid fuels, environmental potentials are mostly negative but differ strongly for PET and PS on the one hand and polyolefins on the

negative potential	positive potential	Global warming impact [kg CO_2-eq]					
		refinery feedstock	gaseous fuels	gasoline	diesel	monomers	chemical upcycling
(2) energy recovery in cement kiln (lignite)	PET TPA			-0.44	-0.41	1.67	2.88
	PET DMT					2.34	
	LDPE	-0.98	-1.44	-0.65	-0.60	-0.21	
	HDPE	-0.99	-1.46	-0.66	-0.60	-0.20	
	PP	-0.83	-1.34	-0.55	-0.50	0.03	
	PS	-0.53	-1.46	-0.73	-0.68	1.43	
negative potential	**positive potential**	**Fossil resource depletion [kg oil eq]**					
		refinery feedstock	gaseous fuels	gasoline	diesel	monomers	chemical upcycling
(2) energy recovery in cement kiln (lignite)	PET TPA			0.34	0.38	1.12	1.57
	PET DMT					1.36	
	LDPE	0.57	0.52	0.67	0.74	0.89	
	HDPE	0.58	0.53	0.68	0.75	0.91	
	PP	0.59	0.53	0.68	0.75	0.97	
	PS	0.64	0.47	0.61	0.67	1.39	

Figure 3.3: Environmental potential for global warming impacts and fossil resource depletion of chemical recycling compared to energy recovery in cement kilns. Red indicates negative environmental potentials. Green indicates positive environmental potentials. White indicates values closer to zero. Grey indicates that chemical recycling does not exist or has been omitted. PET can be used to produce ethylene glycol and two types of monomers: terephthalic acid (PET TPA) and dimethyl terephthalate (PET DMT).

other hand. Environmental potentials of PET (-2.19 kg CO_2-eq) and PS (-1.95 kg CO_2-eq) are clearly negative, while the potentials for polyolefins are close to zero (e.g., 0.01 kg CO_2-eq for HDPE). Mechanical recycling of PET and PS avoids higher global warming impacts by substituting virgin polymer. In contrast, for LDPE, HDPE, and PP, mechanical recycling suffers from significant downcycling (Prognos AG, 2008) as reflected by the substitution factor of 0.7.

Monomer production from plastic packaging waste results mostly in positive environmental potentials if compared to mechanical recycling because substituting the conventional monomers avoids higher global warming impacts. Here, PET achieves the highest environmental potential if dimethyl terephthalate and ethylene glycol (denoted PET DMT in Fig. 1) are produced.

For LDPE, HDPE, and PP, the global warming impact of avoided monomers is lower but still sufficient for positive environmental potentials because downcycling during

mechanical recycling reduces the savings in global warming impacts from virgin polymer production.
Chemical upcycling of PET packaging waste increases the environmental potential of up to 1.13 kg CO_2-eq per kg of plastic packaging waste.
In conclusion, in terms of global warming impacts, mechanical recycling is advantageous to chemical recycling for all plastic packaging wastes if refinery feedstock and fuels are produced. Thus, mechanical recycling should be preferred if possible. In contrast, monomer production and chemical upcycling offer positive environmental potentials if the waste would otherwise be used for mechanical recycling.
For fossil resource depletion, the results slightly differ from global warming: Polyolefins have positive environmental potentials even for chemical recycling to refinery feedstock as well as gaseous and liquid fuels. For polyolefins, the environmental potentials for fossil resource depletion are up to 0.38 kg oil-eq for monomer production. In contrast, for PET, fuel production avoids little fossil resource depletion due to lower net calorific values of PET-derived fuels and, thus, result in negative environmental potentials. Only the monomer production of dimethyl terephthalate and chemical upcycling of PET achieves positive environmental potentials for fossil resource depletion. For PS, qualitatively similar results are obtained for fossil resource depletion and global warming impacts.

3.2.2 Other environmental impacts

Besides global warming and fossil resource depletion, the environmental impacts most commonly assessed for plastic waste management are terrestrial acidification and marine/freshwater eutrophication (Lazarevic et al., 2010). As for global warming and fossil resource depletion, chemical recycling routes with positive and negative environmental potentials for acidification and eutrophication are identified. With respect to the specific routes, the findings for acidification and eutrophication differ from the results obtained for global warming and fossil resource depletion, indicating a potential trade-off in environmental impacts. For instance, chemical recycling of waste currently used in municipal waste incinerators increases terrestrial acidification and freshwater/marine eutrophication impacts if refinery feedstock or fuels are produced. In contrast, all chemical recycling routes have the potential to reduce terrestrial acidification if using plastic packaging waste currently used for energy recovery in cement kilns (cf. Section A.11 in Appendix A).

negative potential	positive potential	Global warming impact [kg CO_2-eq]					
		refinery feedstock	gaseous fuels	gasoline	diesel	monomers	chemical upcycling
(3) mechanical recycling	PET TPA			-2.19	-2.16	-0.08	1.13
	PET DMT					0.59	
	LDPE	-0.38	-0.84	-0.06	0.00	0.39	
	HDPE	-0.42	-0.89	-0.08	-0.03	0.37	
	PP	-0.40	-0.91	-0.12	-0.07	0.46	
	PS	-1.75	-2.68	-1.95	-1.90	0.21	
negative potential	**positive potential**	**Fossil resource depletion [kg oil eq]**					
		refinery feedstock	gaseous fuels	gasoline	diesel	monomers	chemical upcycling
(3) mechanical recycling	PET TPA			-0.86	-0.83	-0.09	0.36
	PET DMT					0.15	
	LDPE	0.04	-0.01	0.14	0.21	0.36	
	HDPE	0.02	-0.03	0.13	0.20	0.35	
	PP	0.00	-0.06	0.09	0.16	0.38	
	PS	-0.43	-0.61	-0.47	-0.40	0.32	

Figure 3.4: Environmental potential for global warming impacts and fossil resource depletion of chemical recycling compared to mechanical recycling. Red indicates negative environmental potentials. Green indicates positive environmental potentials. White indicates values closer to zero. Grey indicates that chemical recycling does not exist or has been omitted. PET can be used to produce ethylene glycol and two types of monomers: terephthalic acid (PET TPA) and dimethyl terephthalate (PET DMT).

3.2.3 Environmental potentials vs. environmental benefits

It is important to emphasize that chemical recycling technologies with positive environmental potentials will not necessarily result in environmental benefits in real-case life cycle assessment assessments, in which data from industrial processes are used for chemical recycling (cf. Section 3.1.4). Additional environmental impacts are generated, for instance, by the separation and purification of chemical products (e.g., TPA and EG), lower conversion rates or additional compounds that must be heated or separated during or before chemical recycling. To understand the influence of additional environmental impacts, a sensitivity analysis for the thermal energy demand and conversion rates of chemical recycling is performed. The focus is on monomer production and chemical upcycling because these achieve positive environmental potentials compared to all benchmark waste treatments (other routes see Section A.12

of Appendix A).
Compared to energy recovery in municipal solid waste incinerators, chemical recycling still has positive environmental potentials for both global warming impact and fossil resource depletion over the complete range of the sensitivity study (cf. Fig. A.8 as well as Fig. 3.5 and A.9 in Appendix A). Thus, plastic packaging waste should most likely not be used in municipal solid waste incinerators but rather be recycled chemically.
For energy recovery in cement kilns, monomer production and chemical upcycling of PET and PS achieve positive environmental potentials for both global warming impacts and fossil resource depletion over the complete parameter range (cf. Figures 3.5 and A.9 in Appendix A). Thus, PET and PS should rather be used for monomer production or chemical upcycling than for any energy recovery option. In contrast to PET and PS, the environmental potential for global warming impacts of PP turns negative if 0.39 MJ additional thermal energy is required, or the conversion rate decreases to 98.5 %. Thus, monomer production of PP probably will not achieve benefits regarding global warming impacts compared to energy recovery in cement kilns in a real-case life cycle assessment.
The comparison of chemical recycling to monomers and mechanical recycling reveals one major challenge (cf. Figure 3.5A): conversion rates have to be higher than a minimum value to achieve any environmental potential for global warming impacts. These minimum conversion rates range from 0.84 for PP to 0.91 for PS (cf. Table 3.2 and Figure 3.5).
The minimal conversion rates of polyolefins and PS are between 21% and 9% higher than the highest reported conversion rates of monomer production. Solely for PET, conversion rates are reported that are approx. 5% higher than the minimal conversion rates. To increase conversion rates, a separation step could be added after the reactor to recycle unconverted polymer back to the reactor. This separation step, however, would require additional process energy (cf. Figure 3.5B). Additional thermal energy would again lower the environmental potential of monomer production compared to mechanical recycling and, thus, increase the minimal conversion rates of chemical recycling: Each additional MJ of thermal energy would increase the minimal process yield by approx. 3% on average for all plastic packaging waste types. In conclusion, monomer production from plastic packaging waste that could be used in mechanical recycling represents a challenging task under more realistic conditions, and further improvements in chemical recycling technologies are needed. Such improvements should increase conversion rates of chemical recycling as well as energy efficiency.

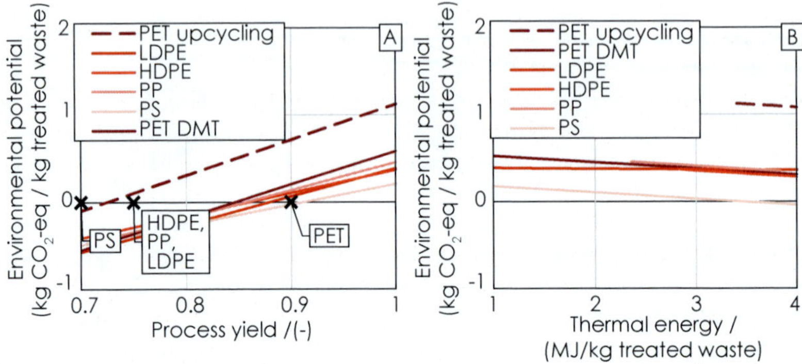

Figure 3.5: Environmental potential for global warming of chemical recycling to monomers and PET upcycling in comparison to mechanical recycling. The values for the environmental potential (y-axis) depend on the process yield (left) or thermal energy demands of chemical recycling (right) (x-axis). The marked values (x) represent the highest reported values of conversion rates for monomer production by chemical recycling. Terephthalic acid production from PET is denoted PET TPA, while dimethyl terephthalate production is denoted PET DMT.

3.2.4 Comparison of results with previous studies

The results of the life cycle assessment study and the sensitivity analysis show the challenge for chemical recycling to compete with the real-case benchmark waste treatment technologies, except waste incineration. Even under best-case conditions (e.g., with minimal environmental impacts), some chemical recycling technologies have higher

Table 3.2: Minimal conversion rates and highest reported values of conversion rates for monomer production of chemical recycling.

parameter	HDPE [1]	LDPE [1]	PP [1]	PET DMT [2]	PS [3]
minimal conversion rates	0.85	0.87	0.84	0.85	0.91
highest reported conversion rates	0.75	0.75	0.75	0.90	0.70

[1] Milne et al. (1999), [2] Paszun and Spychaj (1997), [3] Zhang et al. (1995)

global warming impacts than their benchmark waste treatments. These results are in line with previous life cycle assessment studies for chemical recycling: In 2000, an life cycle assessment study by Patel et al. (2000) found for the treatment of mixed plastic waste that energy recovery in cement kilns, as well as mechanical or chemical recycling reduces CO_2 emission compared to municipal waste incineration. Furthermore, their study suggests that substituting coal, a very carbon-intensive fuel, in cement kilns is advantageous to the substitution of natural gas in other industries. This finding is in line with the present work. The results of Perugini et al. (2005) further confirm that, for the case of PET waste, the combination of mechanical and chemical recycling is beneficial to energy recovery in waste incinerators. They show that mechanical recycling has lower environmental impacts than the combination of mechanical and chemical recycling. The results of these two studies have been confirmed by Lazarevic et al. (2010) in an extensive review of life cycle assessments on plastic waste management scenarios in the context of a future European recycling society. In that review, 77 waste management scenarios were analyzed, including mechanical and chemical recycling but also waste incineration and energy recovery in cement kilns. The review shows a clear preference for mechanical recycling and energy recovery in cement kilns over chemical recycling: for global warming impacts, all scenarios favored mechanical recycling and energy recovery in cement kilns. However, chemical recycling reduced global warming impacts compared to municipal waste incinerators in all scenarios. More recent studies by Maga et al. (2019) underline the benefits of the chemical recycling of packaging waste over waste incinerators, even for new polymer materials like polylactic acid. In contrast to already available literature, this chapter contributes by providing updated inventories for benchmark waste treatments and a comprehensive assessment of 5 major plastic waste streams based on a consistent life cycle assessment methodology applicable to all chemical recycling technologies. This consistent methodology allows the direct comparison across plastic waste streams and, furthermore, to derive robust conclusions about the potential environmental benefits of chemical recycling technologies.

3.3 Conclusions

Recently, chemical recycling technologies have been advocated as enabling technologies for a transition to a circular economy for plastic packaging wastes. To determine whether chemical recycling could lead to environmental benefits, a consistent life cycle assessment method that calculates the maximal environmental benefits by chemical recycling technologies is introduced, denoted the environmental potential of chemical recycling. This method allows us to study the five major plastic packaging wastes PET, HDPE, LDPE, PP, and PS and all currently discussed chemical recycling routes even at early stages of development.

The results suggest that all chemical recycling pathways could reduce global warming impacts and fossil resource depletion if using sorted plastic packaging wastes otherwise treated in municipal solid waste incinerators. Current waste incinerators suffer from high emissions and low efficiencies in producing heat and electricity. The highest potential to reduce global warming impacts and fossil resource depletion is achieved for chemical upcycling of PET to cyclohexane di-methanol instead of its energy recovery in municipal solid waste incinerators: ideal chemical upcycling could avoid up to 4.2 kg CO_2-eq and 1.4 kg oil-eq per 1 kg of treated PET waste. The sensitivity analysis reveals that global warming impacts and fossil resource depletion can still be reduced even for low conversion rates of 70% and high energy demands.

In contrast to using sorted plastic waste treated in municipal solid waste incinerators, plastic waste currently used for energy recovery in cement kilns or mechanical recycling needs a more careful analysis: Energy recovery in cement kilns and mechanical recycling avoids the use of lignite and the production of virgin polymers, respectively. If the objective is to reduce global warming impacts, sorted plastic packaging waste used in cement kilns or mechanical recycling should therefore not be converted to refinery feedstock or fuels. Both products have much lower credits for global warming impacts than substituting lignite or virgin polymers. However, if natural gas or biomass, e.g., waste wood, is used in cement kilns instead of plastic packaging waste, chemical recycling to refinery feedstock and fuels has the potential to reduce environmental impacts.

Compared to refinery feedstock and fuels, chemical recycling of plastic packaging wastes could reduce global warming impacts and fossil resource depletion for most routes to monomers and for upcycling to value-added chemicals. However, a sensitivity analysis shows that environmental benefits regarding global warming impacts of chemical recycling to monomers would still require very energy-efficient processes and increased conversion rates compared to the current state of the art.

Considering terrestrial acidification as well as freshwater and marine eutrophication highlights the need to assess several environmental impacts to understand the full environmental consequences of implementing chemical recycling. For instance, chemical recycling of waste currently used in municipal waste incinerators increases terrestrial acidification and freshwater/marine eutrophication impacts if refinery feedstock or fuels are produced. In contrast, all chemical recycling routes have the potential to reduce terrestrial acidification if recycling plastic packaging waste currently used for energy recovery in cement kilns (cf. Section A.15 of Appendix A).

Even though trends often differ for the five assessed environmental impacts, some general conclusions can be drawn: in particular, (1) PET and PS treated in municipal waste incinerators should preferably be used for monomer production and chemical upcycling as all environmental impacts could potentially be reduced; in contrast, (2) PET and PS currently recycled mechanically should not be used for refinery feedstock or fuel production as all environmental impacts would increase. All other chemical recycling pathways show trade-offs between environmental impacts. The results clearly highlight that choosing a technology solely to reduce global warming impacts might increase other environmental impacts. However, as an overall trend, chemical recycling to monomers and chemical upcycling to value-added products tends to have the potential to reduce more environmental impacts than recycling to refinery feedstock and fuels.

The study analyzed 75 waste treatment scenarios representing approximately 50% of the global plastic waste based on a consistent life cycle assessment methodology. The analysis allows excluding pathways offering no potential to reduce global warming impacts and to match waste types to chemical recycling technologies offering potential benefits. Thus, stakeholders of chemical recycling are provided with the possibility to identify the most promising pathways at an early-stage of development based on a consistent life cycle assessment methodology. However, full life cycle assessment studies are needed for all promising pathways because previous life cycle assessment studies highlighted that it would be very challenging for chemical recycling to compete with energy recovery in cement kilns and mechanical recycling under more realistic conditions. To conduct these full life cycle assessments, this thesis additionally provided up-to-data and industrial datasets for mechanical recycling as well as waste incineration in municipal waste incinerators and cement kilns.

Chapter 4

Life cycle assessment case-study: Towards CO_2-based rubbers

Carbon capture and utilization technologies aim to reduce fossil resource depletion and greenhouse gas emissions (Aresta et al., 2014) by capturing and converting CO_2 into valuable products that can substitute current fossil-based products. In recent years, carbon capture and utilization technologies have been developed for many products such as fuels, chemicals, or polymers (Artz et al., 2018). While many carbon capture and utilization technologies like fuels and chemicals rely on low-carbon electricity to achieve environmental benefits, the utilization of CO_2 as a C1-building block for polyether carbonate polyols (Langanke et al., 2014) has been shown to lead to environmental benefits already for current grid electricity (von der Assen et al., 2014).

Recently, CO_2-based polyether carbonate polyols have been used as rubber precursors (Lipski and Hopmann, 2017; Alagi et al., 2017; Eceiza et al., 2008). The combination of CO_2 with propylene oxide and either allyl glycidyl ether or maleic anhydride results in cross-linkable polyether carbonate polyols with carbon double bonds as unsaturated moieties (Subhani et al., 2016a,b). The cross-linkable polyether carbonate polyols can be further processed by chain-elongation with isocyanates to cross-linkable CO_2-based polyurethanes. These CO_2-based polyurethanes can be utilized as CO_2-based rubbers in synthetic elastomer products (Lipski and Hopmann, 2017).

Even though environmental benefits might be intuitively expected for CO_2-based rubbers, environmental impact reductions for CO_2-based polyols have only been shown for soft and rigid polyurethane foam applications (von der Assen et al., 2014, 2016). In these cases, the environmental benefits mainly depend on the substituted conventional product, namely fossil-based polyols or polyurethanes. However, CO_2-based rubbers substitute completely different products than polyurethanes, for instance, nitrile - or ethylene/propylene-based rubbers. Thus, the environmental benefits of CO_2-based rubbers cannot be taken for granted. To ensure reduced environmental impacts by CO_2-based rubbers in comparison to conventional rubbers, a sound life cycle assessment is needed.

In this Chapter, the environmental impacts of CO_2-based rubbers are compared with their conventional benchmark rubbers. Based on the results, this thesis highlights that a system-wide perspective with cradle-to-grave system boundaries is needed to fully evaluate plastic materials and their environmental impacts. Based on the obtained results, this thesis further shows that large-scale reductions of greenhouse gas emissions are possible but do not necessarily lead to net-zero emission plastic products.

Section 4.1 describes the underlying life cycle assessment methodology and the used datasets. Life cycle assessment results are presented and discussed in Section 4.2. Finally, Section 4.3 summarizes the conclusions about the environmental benefits and disadvantages of CO_2-based rubbers.

Major parts of this Chapter are reproduced with permission of The Royal Society of Chemistry from:

Meys, R., Kätelhön, and Bardow, A. (2019). Towards sustainable elastomers from CO2: life cycle assessment of carbon capture and utilization for rubbers. *Green Chemistry*, 21(12):3334-3342.

The author of this thesis contributed to the development of the concept of the study (including the Supporting Information, cf. Appendix B), performed the life cycle assessment modeling, wrote the first draft of the manuscript, and contributed to the interpretation of the results.

4.1 Life cycle assessment methodology and data

In this section, the life cycle assessment methodology and used datasets are described in further detail.

4.1.1 Goal and scope of the life cycle assessment

The main goal of the presented life cycle assessment is to compare the environmental impact of two types of rubbers: (1) CO_2-based rubbers and (2) conventional rubbers. For this purpose, a cradle-to-grave system boundary including all processes to produce both rubbers is applied in this thesis. In a comparative life cycle assessment, equal life cycle stages or process steps cancel each other out and can be neglected for comparison (Jung et al., 2013). For both CO_2-based and conventional rubbers, similar elastomer recipes result in similar elastomer properties (Lipski and Hopmann, 2017). In consequence, the environmental impacts of the use phases of both rubbers are expected to be equal and therefore are neglected. However, the incineration of both rubbers has to be included in the system boundaries because environmental impacts due to incineration cannot be assumed equal due to different chemical compositions. Section 4.1.3 includes a more detailed description of the compared production systems.

This thesis assesses the following impact categories according to the Recipe Midpoint Hierachist 2016 method (Huijbregts et al., 2016): global warming impact, fossil resource depletion, freshwater, and marine eutrophication, ionizing radiation, stratospheric ozone depletion, particulate matter formation, photochemical ozone formation, and terrestrial acidification. However, primarily the global warming impact and fossil resource depletion are shown because they represent the major objectives of CO_2 utilization (Artz et al., 2018). Furthermore, note that the global warming impact category is measured in CO_2-equivalent emissions (CO_2-eq), which equals the total amount of greenhouse gas emissions. As a result, if this thesis refers to global warming impacts, it is equal to the accumulated sum of all greenhouse gas emissions.

As an example of increased environmental impacts, only ionizing radiation is shown in this chapter since it shows trade-offs with the greenhouse gas emissions. All other results are presented in Appendix B.3. The life cycle assessment software GABI is used for result calculation if not stated otherwise. Datasets for Germany are used where possible because the pilot plant to produce cross-linkable polyether carbonate polyols is located in Germany. In some cases, European datasets are used as a proxy because no German datasets are available.

4.1.2 Functional unit

In life cycle assessment, the functional unit quantifies all functions of the investigated product system (Baumann and Tillman, 2004). Based on the chosen cradle-to-grave approach, the CO_2 system (Figure 4.1) has three functions: (1) the production of the CO_2-based rubber, (2) the production of the side-product of the CO_2 source, ammonia, and (3) the incineration of the CO_2-based rubber. Thus, the system under study is multifunctional.

To solve the multi-functionality problem of the CO_2-source that produces CO_2 and ammonia, this thesis uses the system expansion approach following the recommendations of the ISO standard (ISO 14040, 2021) and recently published guidelines for life cycle assessment of carbon capture and utilization systems (Müller et al., 2020a). In the system expansion approach, the functional unit is expanded to include all functions of a production system. As a result, no product-specific footprints can be obtained, but a transparent picture of the whole production system is provided. The function of the expanded CO_2-based system is, thus, not only the production of CO_2-based rubbers but also the co-production of ammonia at the CO_2 source. To ensure a consistent comparison, the CO_2-based system can only be compared to systems producing the same products. Therefore, the conventional production of ammonia is added to the conventional rubber production system.

For heat and electricity produced by incineration, this thesis uses the so-called cut-off approach (Ekvall and Tillman, 1997). The cut-off approach allocates all environmental impacts of incineration to the rubber production system and is, thus, conservative. As a result, both products are not included in the functional unit.

However, other options to solve the multi-functionality problems might alter life cycle assessment results. Thus, this thesis includes a sensitivity analysis of options to solve multi-functionality on the derived life cycle assessment results and derived product-specific impacts for the CO_2-based rubbers. To derive product-specific footprints, ambiguous choices are required to allocate environmental impacts to either ammonia or the CO_2-based rubbers. To highlight the influence of the allocation procedures for the product-specific environmental impacts (cf. Appendix B.4), this thesis performs a worst-case allocation that allocates all impacts to the CO_2-based rubber and a best-case allocation that uses a credit for produced ammonia, a so-called avoided burden (ILCD, 2010a). For the products of incineration (cf. Appendix B.5), heat and electricity, this thesis also calculates global warming impacts and fossil resource depletion for an avoided burden for heat and electricity. The sensitivity analysis shows that there are no qualitative changes in the life cycle assessment results. For instance, quantitative changes due to the allocation procedure for ammonia vary between 4.83 kg CO_2-eq for the worst-case and 4.57 kg CO_2-eq for the best-case allocation (cf. Appendix B.4).

Finally, the system expansion approach leads to the following functional unit: The CO_2-based rubbers are quantified by the production of 1 kg. To account for the corresponding ammonia production, this thesis assumes the amount of ammonia that is co-produced when utilizing the maximal amount of CO_2 that can be incorporated into the rubber. To ensure the desired product properties, no more than 30 wt. % can be used in the CO_2-based rubbers. Correspondingly, a maximum of 0.185 kg of ammonia is produced as a by-product. Using the maximal amount of ammonia in the functional unit allows comparing conventional to CO_2-based rubbers that incorporate varying amounts of CO_2 (cf. Section 4.2.2). For CO_2-based rubbers with less than 30 wt. % of CO_2, the amount of ammonia co-produced is lower than the amount specified by the functional unit. In this case, the missing amount of ammonia is assumed to be produced by a conventional ammonia plant (cf. Section 4.1.3.1) to ensure the same functional unit. The disposal is accounted for based on the incineration of 1 kg of CO_2-based or conventional rubber in a municipal waste incinerator.

4.1.3 System boundaries and data inventory

In general, the system boundaries include all stages from cradle-to-grave, e.g., production, use phase, and end-of-life (Baumann and Tillman, 2004). Thus, all environmental impacts of upstream processes are included that supply raw materials or utilities (steam, electricity, etc.). However, environmental impacts of use-phases of CO_2-based and conventional rubbers are assumed equal and, therefore, can be neglected in comparative life cycle assessment (Jung et al., 2013) (cf. Section 4.1.1).

4.1.3.1 Production system of CO_2-based rubbers

The system boundary to produce CO_2-based rubbers includes four main parts (cf. Figure 4.1): (1) the CO_2 source, (2) the polyol production, (3) the chain extension, and (4) the waste incineration as end-of-life treatment. The system is referred to as the carbon capture and utilization production system.

The CO_2 source consists of the ammonia plant, CO_2 capture, CO_2 compression, and CO_2 transport to the polyol production plant. The modeling of the ammonia plant is based on a model provided by ecoinvent (Althaus et al., 2007). In this model, ammonia is produced from nitrogen in air and hydrogen from steam methane reform-

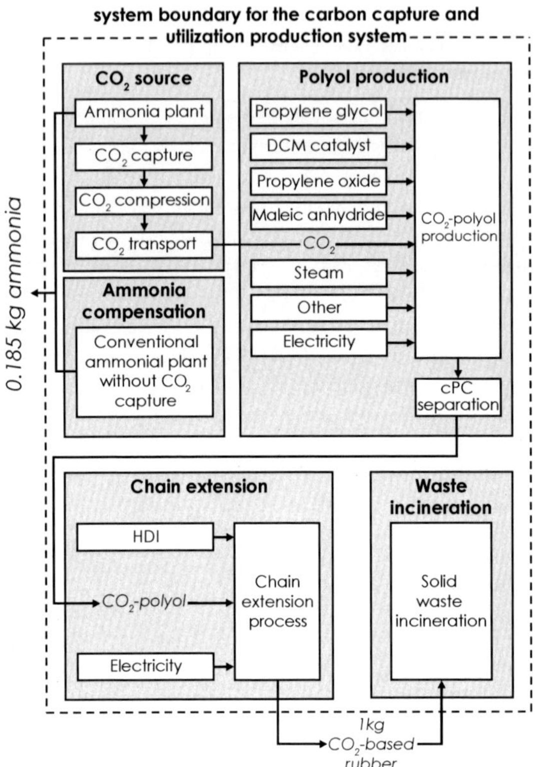

Figure 4.1: The system boundaries for the production of CO_2-based rubbers comprise (1) the CO_2-source, (2) the polyol production, (3) the chain extension as well as (4) the waste incineration as final disposal options. Furthermore, environmental impacts are included for producing raw materials as well as energy carriers and auxiliaries. CO_2 is provided from an ammonia plant. The supply of CO_2 considers capture, compression and transport. The ammonia compensation allows to produce ammonia without utilizing CO_2 for polyol production and thus to keep the amount of ammonia constant for varying amounts of utilized CO_2. Cyclic Propylenecarbonat is denoted cPc in the system boundary.

ing. Ammonia plants already include a CO_2 capture if hydrogen from steam methane reforming is used because the CO_2 acts as catalyst poison in the ammonia reactor (Farla et al., 1995). The CO_2 capture in the used model is based on the monoethanol amine absorption process that is already included in the ecoinvent dataset (Althaus et al., 2007). Additional 0.4 kWh electricity is needed to pressurize CO_2 for storage at 100 bar (Farla et al., 1995). Subsequently, the pressurized CO_2 is transported to the polyol production plant. Furthermore, an optional ammonia compensation is included to account for the difference between the actual amount of ammonia produced and the maximal amount of ammonia produced. For instance, only 0.134 kg of ammonia is co-produced if CO_2-based rubbers include 20 wt. % CO_2. The ammonia compensation, thus, produces 0.051 kg ammonia to provide the overall 0.185 kg of ammonia in the functional unit. It should be noted that ammonia production is a preferred source of CO_2 since CO_2 is separated anyhow (Müller et al., 2020b).

The primary feedstock demand data for CO_2, propylene oxide, maleic anhydride, and the catalyst were collected from a pilot plant that produces cross-linkable polyether carbonate polyols. Propylene oxide is modeled according to its production mix: 43 % chlorohydrin process, 27 % production process with styrene monomer as co-product, 16 % production process with t-butyl alcohol as co-product, 10 % hydrogen peroxide process and 4 % Sumitomo process with cumene hydroperoxide as oxidant (Chem-Systems, 2009). However, the Sumitomo process was neglected due to missing life cycle assessment data. Maleic anhydride is modeled according to the production process based on n-butane (Sphera, 2019). Catalyst production, energy requirements for steam and electricity, as well as other utilities in polyol production are based on von der Assen et al. (2014).

During the polyol production, cyclic propylene carbonate (cPc) is formed as a by-product. CPc yields between 0.7 and 1.7 % of mass input have been reported (Subhani et al., 2016a). This thesis assumes a conservative value of approx. 4 % yield to cPc. The side product cPc is separated from the actual CO_2-based polyol by a thin-film evaporator. The energy demand for evaporation of the side product is based on the enthalpy of evaporation of cPc. The side product cPc is allocated according to its mass. Past studies showed that changes in the allocation procedure for cPc only slightly influence life cycle assessment results for polyol production between -0.3 to 3.8 % (von der Assen et al., 2014). Thus, only mass allocation is used in this thesis. No environmental impact of plant construction is considered because the environmental impacts of chemical plant construction are typically very small in the chemical industry (Frischknecht et al., 2007) and probably similar for CO_2-based and conventional rubber.

The chain extension step is based on the chain elongation in which CO_2-based polyols are mixed with hexamethylene diisocyanate (HDI) in an extruder. The electricity demand of extrusion is based on measurements at a pilot-scale plant. During the chain extension process, the CO_2-based polyols and HDI are completely converted to CO_2-based rubbers.

The incineration is modeled according to Doka (Doka, 2003, 2013). This model includes all environmental impacts of flue gas emissions, flue gas cleaning, and the disposal of residuals. All emissions are allocated to the waste treatment and not to any potential energy production in the form of steam, heat, or electricity (Doka, 2003, 2013).

4.1.3.2 Production system of conventional rubbers

The system boundary of producing conventional rubbers includes three major steps (Figure 4.2): (1) the ammonia compensation, (2) the conventional synthetic rubber production, and (3) the incineration as end-of-life treatment. The system is referred to as the conventional production system. The ammonia plant is modeled with the same model provided by ecoinvent as in the carbon capture and utilization production system. However, CO_2 is assumed to be released into the atmosphere instead of its utilization. This ammonia compensation produces 0.185 kg of ammonia according to the defined functional unit (cf. Section 4.1.2)

The production of conventional rubbers includes all environmental impacts of raw material and utility supply, as well as all direct emissions from rubber production. This thesis assesses four types of conventional rubbers for comparison: (1) hydrogenated nitrile butadiene rubber (HNBR), (2) nitrile butadiene rubber (NBR), (3) ethylene propylene diene rubber (EPDM), and (4) polychloroprene (CR). These four types of rubber products have been shown to result in very similar properties of final elastomer products as the CO_2-based rubbers. All datasets are based on full life cycle assessment datasets provided in the GABI software (Sphera, 2019). In general, the Gabi database is a well-accepted database for life cycle assessment data with overall good data quality. According to the data quality rating of the GABI software, all datasets used in this thesis have an overall representation of the reality of good or very good (Sphera, 2019).

HNBR is produced by the hydrogenation of NBR with pure hydrogen (Sphera, 2019; Happ et al., 2011). NBR is produced by the co-polymerization of acrylonitrile and butadiene. For the hydrogenation, NBR is first dissolved using toluene as a solvent in a solution production step. The solution is fed into a reactor, and hydrogen is added under high pressure of approx. 50 bar. In the presence of a heterogeneous

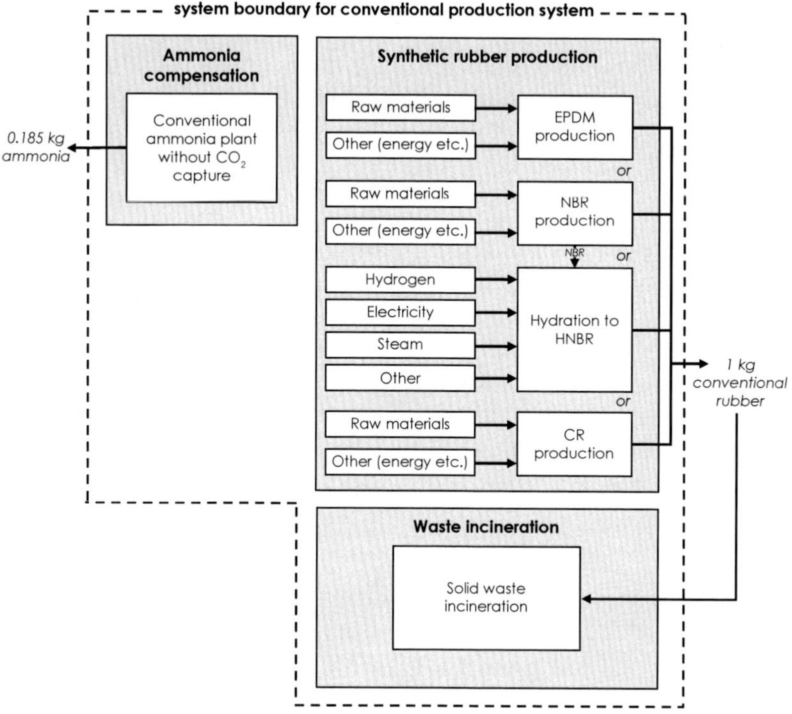

Figure 4.2: The system boundaries for the production of conventional rubbers comprise (1) the ammonia plant without CO_2 utilization (ammonia compensation), (2) the conventional rubber production as well as (3) the incineration. Furthermore, environmental impacts are included for producing raw materials as well as energy carriers and auxiliaries.

catalyst, NBR and hydrogen react to HNBR. Subsequently, the catalyst is removed by filtration and recycled back into the hydrogenation reactor. Finally, HNBR is isolated by direct evaporation of toluene. Toluene is recycled into the solution production step. NBR and HNBR can be produced with different acrylonitrile contents that strongly influence the material properties of elastomer products. Typical acrylonitrile contents range from 15 % to 50 % (Elvers and Ullmann, 2011). This thesis considers an average acrylonitrile content of 33 % (Sphera, 2019).

EPDM is produced by the copolymerization of ethylene, propylene, and a diene, here: norbornene (Sphera, 2019). The process is operated at 10 to 50 °C and 1 to 20 bar in a continuously stirred tank reactor. The exothermal reaction is cooled. Typically, solvents like hexane are used. After the reactor, unreacted resources and solvents are recycled through a separation step. Finally, the product is dried with hot air.

CR is produced by emulsion polymerization of chloroprene (CR) that is produced from butadiene in three steps: First, butadiene is chlorinated by heating up to 250 °C and pressures of 1-7 bar in order to form a mixture of di-chlorobutene isomers. Second, the di-chlorobutene mixture is isomerized in the presence of a catalyst at temperatures from 60-120 °C. Third, after dehydrochlorination of the mixture with diluted NaOH between 40 and 80 °C, chloroprene is obtained and purified.

After the use phase, each conventional rubber is incinerated in a municipal solid waste incinerator. As for the carbon capture and utilization production system, the incineration is modeled according to Doka (Doka, 2003, 2013).

4.2 Life cycle assessment results and discussion

In the following sections, the life cycle assessment results are presented and discussed.

4.2.1 Results for the system-wide assessment

In this section, results are presented for the functional unit of 1 kg of rubber, 0.185 kg of ammonia, and the incineration of 1 kg of rubber as a component of a final elastomer (cf. Section 4.1.2). The results are shown in detail for the comparison of CO_2-based rubbers to the conventional rubber HNBR (Figure 4.3). HNBR is the conventional rubber where emissions are reduced most, but environmental impacts are qualitatively similar for all conventional rubbers. All results are presented for the following recipe for cross-linkable polyether carbonate polyols: approx. 20 wt. % CO_2, approx. 2 % double bond moieties (maleic diester) and a molar mass of polyether carbonate polyols of approx. 4200 g/mol. This recipe results in CO_2-based rubber properties that allow their utilization in synthetic elastomer products such as sealants and flexible tubes.

Major contributions to the global warming impact of conventional rubbers result from rubber production and incineration. For HNBR, the rubber production leads to 53.5 % and the incineration to 41.8 % of the total global warming impact per functional unit. The compensation for ammonia only contributes with 4.8 %. In the case of the

CO_2-based rubbers, major contributions to global warming impacts result from polyol production (50.3 %) and the chain extension (6.4 %) step. Thus, CO_2-based rubber production represents the major share of 56.7 % of the total global warming impact. Incineration of CO_2-based rubbers contributes 39.8 % to the total global warming

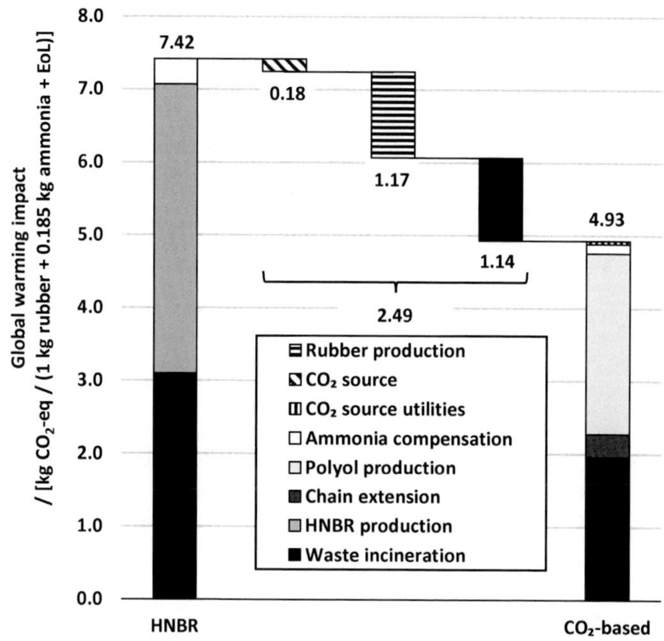

Figure 4.3: Global warming impacts in kg CO_2-equivalents of cradle-to-grave product systems for hydrogenated nitrile butadiene rubber (HNBR) and CO_2-based rubber. The CO_2 source utilities include additional CO_2 compression, and transport. The shown CO_2 source reductions equal the global warming impacts of ammonia compensation in the conventional production system minus those of ammonia compensation and CO_2 source utilities in the carbon capture and utilization production system. Rubber production reductions are the difference of global warming impacts from the rubber production of conventional and CO_2-based rubbers (polyol production and chain extension). The waste incineration reductions equal the difference of waste incineration impacts of both rubbers.

impact.

The overall global warming impacts are reduced by 34 % for the comparison of HNBR with CO_2-based rubber. This reduction results partly from the CO_2 source (11 %), but the major reductions are due to the rubber production (43 %) and the difference in waste incineration (46 %). Reductions at the CO_2 source are due to avoided CO_2 emissions from the ammonia plant minus the additional global warming impacts from utility supply for CO_2 compression and transport (CO_2 source utilities). Reductions in rubber production are based on the difference between conventional rubber production and CO_2-based rubber production (polyol production plus chain extension). This reduction is mainly based on the exchange of raw materials of conventional rubbers by CO_2 and propylene oxide. The global warming impact reduction of waste incineration is mainly due to the lower carbon content of the CO_2-based rubbers compared to HNBR. CO_2-based rubbers have a carbon content of 53.5 wt. %, while HNBR has a carbon content of 79.9 wt. % (cf. Appendix B.2). Thus, HNBR results in higher global warming impacts than CO_2-based rubber.

The overall global warming impact for all conventional and CO_2-based rubbers is shown in Figure 4.4. CO_2-based rubbers reduce global warming impacts compared to all conventional rubbers. The reductions range from 18 % for CR up to 34 % for HNBR.

Former publications highlighted the need to distinguish between the amount of CO_2 utilized and avoided CO_2-eq emissions (von der Assen et al., 2013). These publications showed that a specific reduction of 2.98 kg of CO_2-eq emissions can be achieved if 1 kg of CO_2 is utilized to produce polyols for polyurethane soft foams (von der Assen et al., 2014). Figure 4.5 shows the specific global warming impact reductions per 1 kg of utilized CO_2 if conventional rubbers are substituted by CO_2-based rubbers.

In all cases, the reduced global warming impacts per utilized kg of CO_2 are higher than the amount of utilized CO_2. The reduction per amount of utilized CO_2 equals up to 12 kg of CO_2-eq if HNBR is substituted by CO_2-based rubbers. Thus, the global warming impact reductions per 1 kg of utilized CO_2 for CO_2-based rubbers can be more than four times larger than the value previously reported for polyurethane soft foams (von der Assen et al., 2014).

In Figure 4.6, CO_2-based rubbers are compared with conventional rubbers with regard to fossil resource depletion. Fossil resource depletion is mainly reduced by the substitution of conventional rubber production by CO_2-based rubbers. In contrast to reducing global warming impacts, incineration has nearly no effect on reducing fossil resource depletion. Overall, fossil resource depletion is reduced from 17 % for CR and up to 33 % for HNBR.

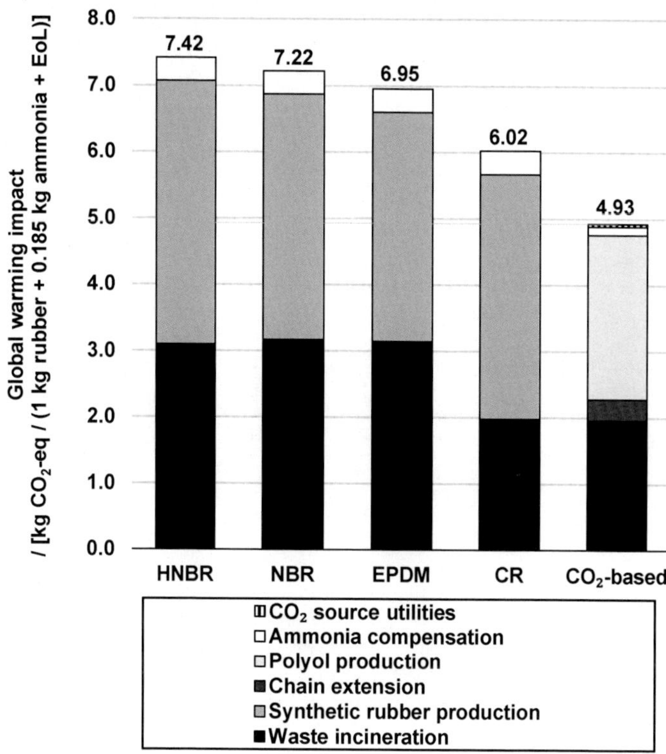

Figure 4.4: Global warming impacts in kg CO_2-equivalents of cradle-to-grave product systems for conventional and CO_2-based rubbers. The CO_2 source utilities include CO_2 compression and CO_2 transport. Four conventional rubbers are shown: (1) hydrogenated nitrile butadiene rubber (HNBR), (2) nitrile butadiene rubber (NBR), (3) ethylene propylene diene rubber (EPDM) and (4) polychloroprene (CR).

The results for the other assessed impact categories differ from those of the global warming impact and fossil resource depletion and, thus, are highlighted in the following. While CO_2-based rubbers reduce photochemical ozone depletion compared to all conventional rubbers, other impacts are increased in some cases. In the following, these increased environmental impacts are described briefly. Ionizing radiation

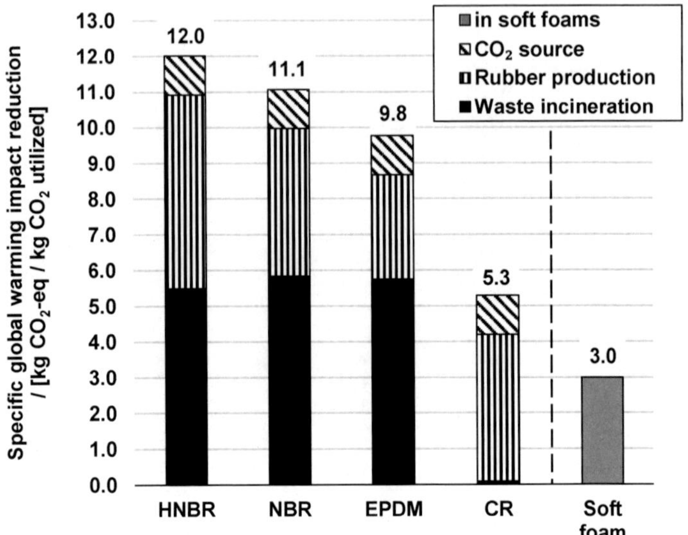

Figure 4.5: Specific global warming impact reductions in kg CO_2-equivalents per 1 kg of CO_2 utilized in CO_2-based-based rubbers. Additionally, the global warming impact reduction is shown for the utilization of CO_2-based-based polyols in polyurethane soft foam applications (von der Assen et al., 2014). Four conventional rubbers are shown: (1) hydrogenated nitrile butadiene rubber (HNBR), (2) nitrile butadiene rubber (NBR), (3) ethylene propylene diene rubber (EPDM) and (4) polychloroprene (CR).

is increased between 52 % compared to HNBR to 53 % compared to EPDM. Ionizing radiation can be attributed to emissions from the use of nuclear fuels or the burning of coal for electricity production. The electricity demand is higher for the production of propylene oxide than for HNBR and NBR due to energy-intensive processes (e.g. the HPPO process; or chlor-alkali electrolysis in the chlorohydrin processes). Thereby, propylene oxide supply contributes 67 % to the ionizing radiation of CO_2-based rubbers (Figure 4.7). However, the ionizing radiation impact to produce 1 kWh of electricity is lower for wind or solar power plants than for nuclear power plants (Sphera, 2019). Thus, ionizing radiation is expected to decrease with the transition to renewable electricity in Germany. However, currently, the utilization of propylene oxide is the main reason why, in some cases, CO_2-based rubbers increase environ-

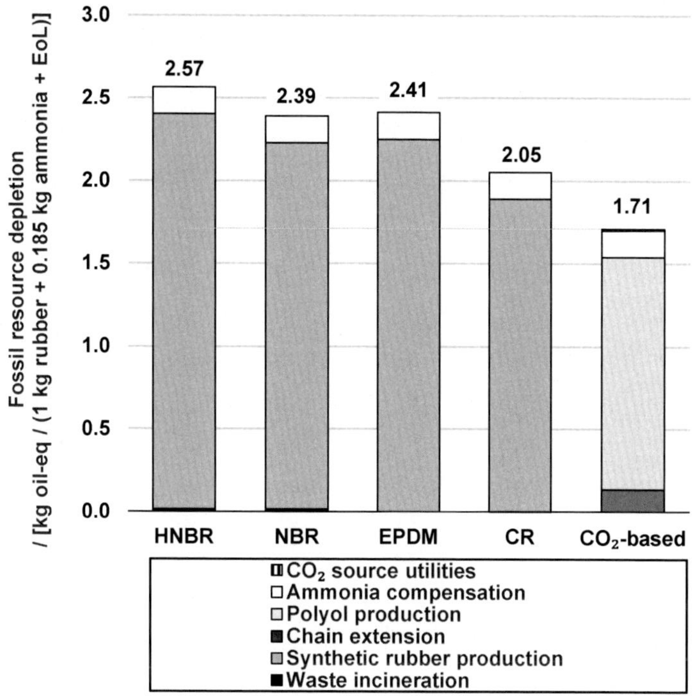

Figure 4.6: Fossil resource depletion in kg oil-equivalents of cradle-to-grave product systems for conventional and CO_2-based rubbers. The CO_2 source utilities include CO_2 compression and CO_2 transport. Four conventional rubbers are shown: (1) hydrogenated nitrile butadiene rubber (HNBR), (2) nitrile butadiene rubber (NBR), (3) ethylene propylene diene rubber (EPDM) and (4) polychloroprene (CR).

mental impacts. In case of particulate matter formation, CO_2-based rubbers result in an increase of 18 % compared to EPDM, but decrease impacts between 25 % for NBR and 30 % for HNBR. For stratospheric ozone depletion, environmental impacts of CO_2-based rubbers are increased compared to EPDM (38 %) and CR (1 %). However, stratospheric ozone depletion is decreased substantially compared to NBR (371 %) and HNBR (364 %). For terrestrial acidification, CO_2-based rubbers increase environmental impacts compared to EPDM (26 %) and decrease impacts between 25 % for CR up to 38 % for HNBR. For freshwater eutrophication, environmental impacts compared to all conventional rubbers are increased between 12 % for CR up to 30 % for EPDM.

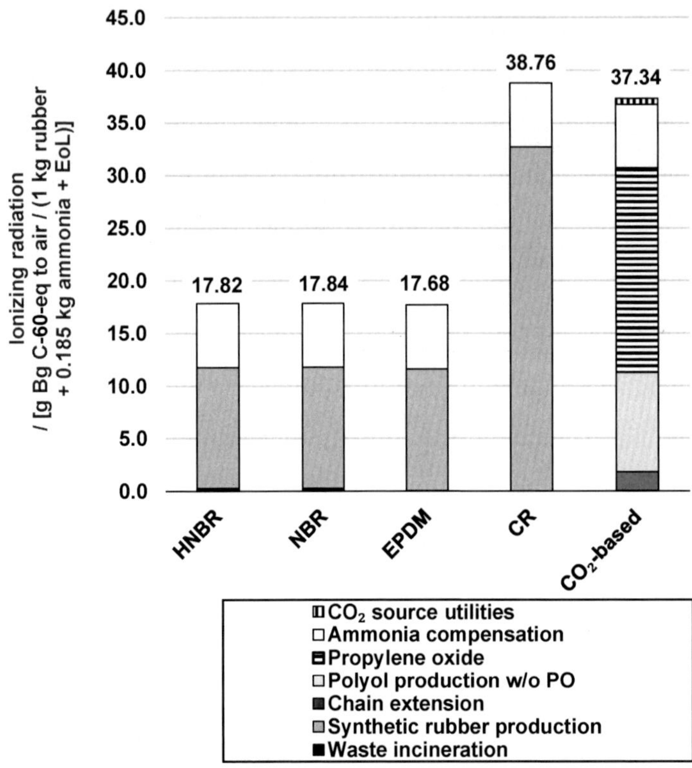

Figure 4.7: Ionizing radiation in kg Bg C60-equivalents to air of cradle-to-grave product systems for conventional and CO_2-based rubbers. The "polyol production w/o PO" includes environmental impacts from all raw materials and energy carriers used for polyol production except propylene oxide (PO) which is shown separately. The CO_2 source utilities include CO_2 compression and CO_2 transport. Four conventional rubbers are shown: (1) hydrogenated nitrile butadiene rubber (HNBR), (2) nitrile butadiene rubber (NBR), (3) ethylene propylene diene rubber (EPDM) and (4) polychloroprene (CR).

In the case of marine eutrophication, environmental impacts are increased compared to EPDM (55 %) and CR (7 %) but are substantially decreased compared to NBR (334 %) and HNBR (326 %). More detailed contributions to the other environmental impacts can be found in Appendix B.3. From the results for other environmental impacts, it can be seen that a holistic assessment of several environmental impacts is needed in order to understand the potential burden shifting from one environmental impact to another.

4.2.2 Varying recipes of CO_2-based rubbers

The standard recipe for CO_2-based rubbers in Section 4.2.1 is fit to sealant and flexible tube elastomer products (Lipski and Hopmann, 2017). However, other elastomer products might require a variation of CO_2-based rubber recipes to achieve a different set of elastomer properties. Thus, it seems useful to assess the changes of environmental impacts due to changes in CO_2-based rubber recipes. For this purpose, the global warming impact for CO_2-based rubbers is computed as a function of the CO_2-content and the molar mass of cross-linkable polyether carbonate polyols (Figure 4.8). Experimental values ranging from 0 to 30 wt. % of CO_2 and 500 to 5500 g/mol are used for the calculations.

It can be observed that low molar masses and low CO_2-contents of polyols increase the environmental impacts of CO_2-based rubbers from 4.34 kg CO_2-eq (upper right corner in Figure 4.8) up to 7.03 kg CO_2-eq (lower-left corner of Figure 4.8). Thus, global warming impacts could be further reduced compared to the life cycle assessment results in Section 3.1 by using higher CO_2-contents and higher molar masses of polyols. The maximal reductions of the global warmings impacts range from 26 % for CR to 40 % for HNBR if a polyol with 5500 g/mol and a CO_2-content of 30 wt. % is used.

Increasing the CO_2-contents decreases global warming impacts because greater fractions of propylene oxide are substituted and its impacts are avoided. The molar mass of polyether carbonate polyols dictates the polyol-to-isocyanate ratio per 1 kg of CO_2-based rubber. If the molar mass is increased, the share of isocyanate in the final CO_2-based rubber is decreased. Hexamethylene diisocyanate has higher global warming impacts per kg than polyether carbonate polyol. Thus, by increasing the molecular weight and thus decreasing the isocyanate share, the overall global warming impact per functional unit is decreased.

Based on the variations of product properties, threshold values can be obtained that represent the minimal molar masses and CO_2-contents necessary to reduce global warming impacts compared to conventional rubbers. For instance, the line for 6.02 kg

CO_2-eq per functional unit represents the environmental impacts of CR. Thus, the 6.02 kg CO_2-eq solid line can be used to obtain minimal CO_2-contents for a certain molar mass of polyether carbonate polyols to reduce global warming impacts compared to CR.

The results show that purely fossil-based polyurethane rubbers (CO_2-content of 0 wt. %) do not reduce global warming impacts compared to CR. For a molar mass of 1000 g/mol, a minimal CO_2-content of 12.2 wt. % has to be used if CO_2-based rubbers should reduce global warming impacts compared to CR. Thus, only by utilizing sufficient CO_2, global warming impacts are reduced compared to all conventional rubbers. However, for NBR and HNBR, all recipes of CO_2-based rubbers reduce global warming impacts.

4.3 Conclusion and discussion

The presented life cycle assessment for the comparison between CO_2-based and conventional rubbers shows that CO_2-based rubbers with approx. 20 wt. % CO_2 could reduce global warming impacts by 1.1 kg CO_2-eq compared to CR up to 2.49 kg CO_2-eq compared to HNBR. The specific global warming impact reduction ranges from 5.33 to 12 kg CO_2-eq emissions per 1 kg of utilized CO_2 for CR and HNBR, respectively. In the case of fossil resource depletion, environmental impacts are reduced between 0.34 kg oil-eq compared to CR up to 0.85 kg oil-eq compared to HNBR. In contrast to the latter reductions, CO_2-based rubbers increase other environmental impacts such as stratospheric ozone depletion or marine eutrophication between 1 and 54 %, respectively. However, for stratospheric ozone depletion, environmental impacts are decreased substantially compared to NBR (371 %) and HNBR (364 %).

The global warming impacts are mainly reduced by the substitution of conventional rubbers and reduced emissions from end-of-life waste incineration. In case of the fossil resource depletion, the largest reductions are due to the substitution of conventional rubbers. Furthermore, environmental benefits strongly depend on the chemical recipe of the CO_2-based rubbers. In particular, the CO_2-content and molar mass of polyether carbonate polyols influence the environmental impacts of CO_2-based rubbers. For instance, CO_2-based rubbers have higher global warming impacts than CR for a molar mass of polyols equal to 1000 g/mol and if the CO_2 content drops below 12.2 wt. %. All recipes reduce global warming impacts if HNBR and NBR are substituted by CO_2-based rubbers. Thus, solely by utilizing enough CO_2, global warming impacts reductions are possible compared to all assessed conventional rubbers.

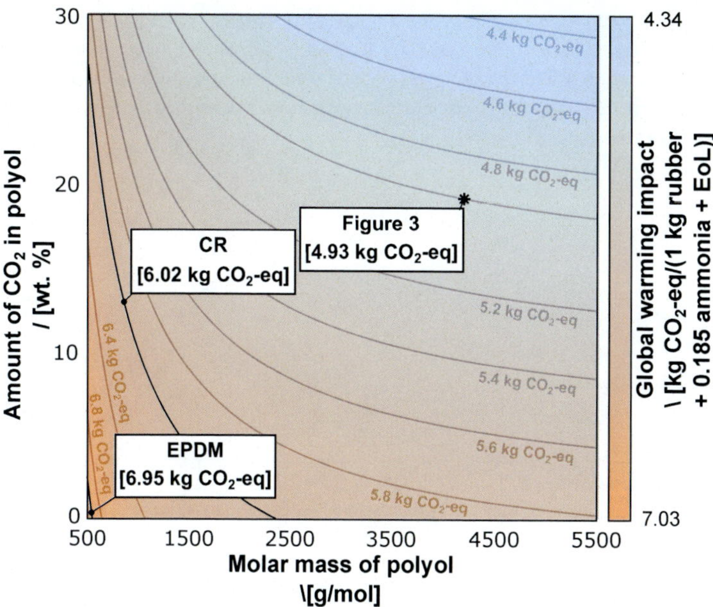

Figure 4.8: Global warming impacts of CO_2-based rubbers as function of the CO_2-content and the molar mass of cross-linkable polyether carbonate polyols. The global warming impact is calculated based on varying CO_2 contents and molar masses. Solid lines indicate constant global warming impacts. Solid black lines represent environmental impacts of conventional rubbers as indicated. Furthermore, the global warming impact of CO_2-based rubbers from Section 4.2.1 is marked. For NBR and HNBR, no threshold values can be calculated, because all recipes for CO_2-based rubbers reduce global warming impacts.

In conclusion, CO_2-based rubbers introduce a novel class of CO_2-based polymers to the market that offers the possibility to reduce global warming impacts and fossil resource depletion compared to conventional rubbers. While per kg of utilized CO_2, up to 12 kg of CO_2 could be reduced, this large-scale reduction does not lead to net-zero emission plastic products. It is important to notice that net-zero claims warrant a holistic life cycle assessment of all relevant life cycle stages, e.g., from cradle-to-grave (Finkbeiner and Bach, 2021). As can be seen from the case of CO_2-based rubbers, large reductions in greenhouse gas emissions do not necessarily lead to net-zero emission plastic materials. Furthermore, reducing greenhouse gas emissions of rubber products via CO_2 utilization might lead to a problem shifting from one impact to another. By this means, in this chapter, the scientific gaps 1 and 2 described in Section 2.4 have been addressed.

Chapter 5

Bottom-up model of plastic life cycles

This chapter highlights how this thesis addresses all scientific gaps described in Section 2.4 by building the first of its kind bottom-up model of plastic production and plastic waste treatment. To build this bottom-up model, this thesis revised over 400 technical datasets such that they are based on a consistent, life cycle assessment compliant methodology. To enable the calculation of greenhouse gas emissions and other environmental impact, the datasets are included in a bottom-up model that is based on the methodology of the Technology Choice Model (TCM) (Kätelhön et al., 2016). The TCM enables the assessment of environmental impacts of large production systems and is suited to incorporate technological, environmental, and economic parameters (Kätelhön et al., 2016). Based on the bottom-up model and since fossil as well as all circular technologies are included, a system-wide and systematic evaluation of net-zero plastics is enabled.

This chapter provides a detailed explanation of the bottom-up model used to represent the life cycle of plastics. First, the general scope of the bottom-up model is described in Section 5.1. Afterward, the computational structure of the bottom-up model is explained in Section 5.2. Section 5.3, gives an explanation of the data sources used in this thesis. The final Section 5.4 explains the methodology to close data gaps based on thermodynamic modeling.

Major parts of this chapter are reproduced with permission of Science from:

Meys, R., Kätelhön, A., Bachmann, M., Winter, B., Zibunas, C., Suh, S. and Bardow, A. (2021). Achieving net-zero greenhouse gas emission plastics by a circular carbon economy. *Science*, 374(6563):71-76.

The author of this thesis developed the concept of the study, was in charge of the development of the set-up of the bottom-up model and methodology, wrote the first draft of the manuscript, and interpreted the results.

5.1 Scope of the bottom-up model

This Section describes the general scope of the bottom-up model by providing information about the included chemicals, plastics, and plastic wastes (Section 5.1.1), as well as the system boundary, the final demand, and allocation principles (Section 5.1.2). A final Section 5.1.3 briefly highlights potential other uses of the model beyond plastics.

5.1.1 Included chemicals, plastics, and plastic wastes

The bottom-up model focuses on the life cycle of plastics. However, over 50 plastic materials are currently commercially used in industry (Alauddin et al., 1995). Therefore, the scope of the bottom-up model has to be balanced between being (1) large enough such that the results are valid and sufficiently representative for the global plastic life cycle and (2) small enough such that the modeling of the plastic life cycle is actually feasible within the scope of this thesis. Taking into account global plastic production (cf. Figure 5.1), 14 plastics were found to represent over 90% of the global plastic production (Geyer et al., 2017). Thus, these 14 plastics are used as the scope of this thesis. To fully represent the life cycle of the 14 plastics, over 75 chemical and polymer intermediates are added to the bottom-up model (cf. next paragraph). These chemical and polymer intermediates are required to represent the production from chemical feedstock to plastic products. As a result, the bottom-up model does not only represent over 90% of global plastics but also at least 75% of the global greenhouse gas emissions and 60% of the energy consumption of the chemical industry (Ausfelder et al., 2013).

Currently, plastic packaging represents over 50% of the global plastic waste volume and packaging waste is the only waste that can be properly recycled mechanically (Geyer et al., 2017; Ragaert et al., 2017). Thus, this thesis subdivides plastic waste into two categories: (1) plastic packaging waste and (2) other plastic wastes. Other plastic wastes are derived from transportation, building and construction, electronics, consumer and institutional products, industrial machinery, and miscellaneous (Geyer et al., 2017).

Considering the 14 plastics and their wastes, the bottom-up model mainly covers chemical structures consisting of carbon. However, the production of the 14 plastics requires several non-carbon chemical intermediates that are not necessarily used as monomers but as auxiliaries. Thus, some non-carbon chemicals, e.g., ammonia or caustic soda, are included in the bottom-up model since they are required during

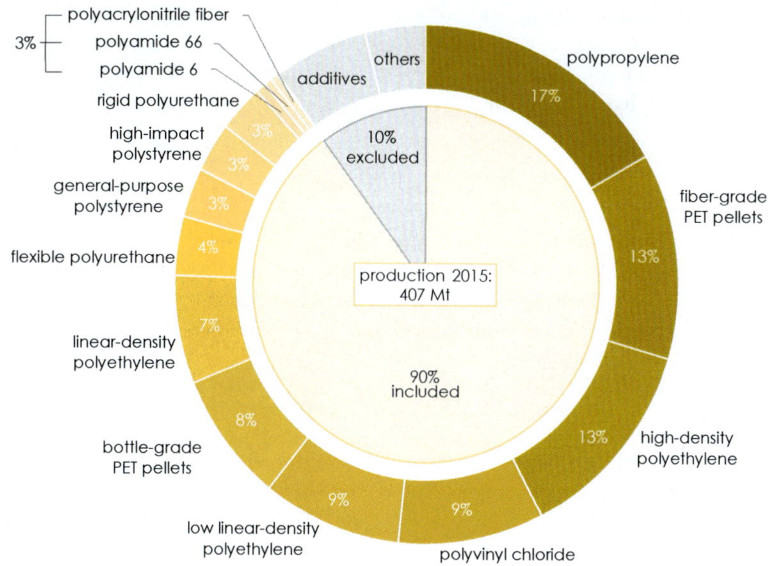

Figure 5.1: Production volume of plastics in 2015 according to Geyer et al. (2017).

plastic production. The following chemicals, plastics, and plastic wastes are covered by the bottom-up model:

Chemicals: acetic acid, acetone, acetonitrile, acrylic acid, acrylonitrile, adipic acid, allyl chloride, ammonia, aniline, benzene, butadiene, C4 fraction, calcium chloride, calcium oxide, caprolactam, carbon dioxide, carbon monoxide, caustic soda (50%), chlorine, cumene, cyclohexane, dichloropropylene, diethylene glycol, dimethyl terephthalate, dinitrotoluene, dipropylene glycol, epichlorohydrin, ethane, ethanol, ethylbenzene, ethylene, ethylene glycol, ethylene oxide, formaldehyde, glycerin, hexamethylenediamine, hydrogen, hydrogen cyanide, methanol, methyl acrylate, methylene diphenyl diisocyanate, monoethanolamine, naphtha, natural gas, nitric acid (60%), nitrobenzene, nitrogen, oleum (33%), oxygen, o-xylene, phenol, polybutadiene, polyester polyol, polyether polyol, propionitrile, propylene, propylene glycol, propylene oxide, p-xylene, pyrolysis gasoline, silicon carbide, sodium carbonate, sodium chloride, styrene, sulfur trioxide, sulfuric acid, synthesis gas (2:1), terephthalic acid, toluene, toluene diisocyanate, vinyl chloride, mixed xylenes.

Plastics: polyamide 6, polyamide 66, fiber-grade PET pellets, bottle-grade PET pellets, polyacrylonitrile fiber, high-density polyethylene, low-density polyethylene, low linear-density polyethylene, polypropylene, general-purpose polystyrene, high-impact polystyrene, flexible polyurethane, rigid polyurethane, polyvinyl chloride.

Plastic packaging wastes: bottle-grade PET pellets, high-density polyethylene, low-density polyethylene, low linear-density polyethylene, polypropylene, general-purpose polystyrene, high-impact polystyrene.

Other plastic wastes: polyamide 6, polyamide 66, fiber-grade PET pellets, bottle-grade PET pellets, polyacrylonitrile fiber, high-density polyethylene, low-density polyethylene, low linear-density polyethylene, polypropylene, general-purpose polystyrene, high-impact polystyrene, flexible polyurethane, rigid polyurethane, polyvinyl chloride.

5.1.2 System boundary, final demand, and allocation

The system boundary depicts all life cycle stages from cradle-to-grave that are included in the bottom-up model (Figure 5.2). The system boundary includes five modules that again comprise multiple production technologies. Three modules represent the supply of carbon feedstock: fossil resources (natural gas, oil, coal), CO_2, and biomass. The fourth module represents the production technologies of all chemicals and plastics. The waste treatment of plastics is included in the fifth module. The use-phase is not considered in the bottom-up model because all pathways produce identical chemicals and plastics and treat the same type of plastic wastes. Thus, environmental impacts of the use-phase would cancel each other in the comparison of two different plastic life cycles, e.g., based on fossil resources or biomass as carbon source. However, the use-phase defines the amount of plastics produced and the respective need for plastic waste treatment (denoted as "final demand"). In the following, each module and the final demand are briefly described. A summary of literature sources and covered technologies for all modules can be found in Appendix C. Furthermore, the author of this thesis provides a comprehensive list of all mass and energy balances that can be made publicly available on Zenodo (Raoul Meys, 2021). Proprietary data from industry partners is not provided.

Modules 1 to 3 - carbon feedstock supply. The carbon feedstock supply includes three options: First, CO_2 supply by capture from ammonia, ethylene oxide, hydrogen-producing plants, biomass fermentation or gasification, waste incinerators, or directly from ambient air (approx. 400 ppm CO_2 concentration) are included. Fossil CO_2 point sources that generate electricity, e.g., gas- and coal-fired power plants, have been excluded since these sources are not expected to be part of the analyzed net-zero

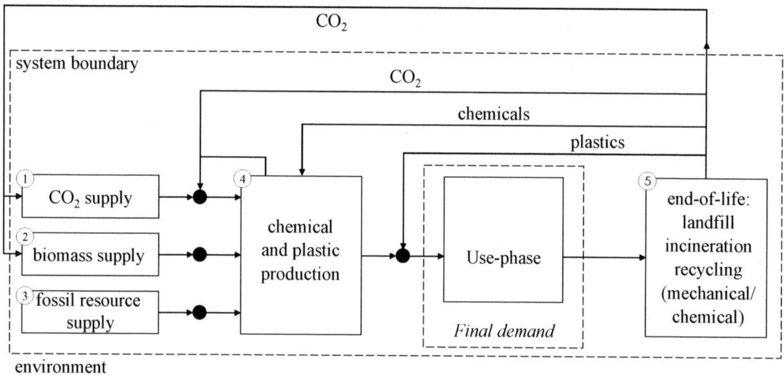

Figure 5.2: Overview of system boundary and modules of the bottom-up model: CO_2 supply, biomass supply, fossil resource supply, chemical and plastic production, and end-of-life. The use-phase is included for clarity reasons but is not part of the system boundary. However, the plastics entering and plastic wastes leaving the use-phase represent the final demand of the model. The end-of-life includes landfill, incineration (i.e., energy recovery), and mechanical and chemical recycling.

future (Müller et al., 2020b). Second, the lignocellulosic biomass feedstock is included, e.g., wood chips, bark chips, wood pellets, or miscanthus. Biomass provision includes all environmental impacts from harvesting and production, e.g., nitrous oxide and methane, but also emissions due to harvesting machinery. Third, fossil resource supply is included for which the environmental impacts include raw material extraction, e.g., crude oil or natural gas, over distribution to downstream processing (Althaus et al., 2007) up to the final generation of the fossil feedstock, e.g., naphtha. In summary, the carbon feedstock supply includes all environmental impacts from cradle, e.g., the natural environment, to the gate, e.g., the supply of one kg of fossil feedstock, CO_2, or biomass.

Module 4 - chemical and plastic production. During chemical and plastic production, fossil resources, chemical intermediates from chemical recycling, plastics from mechanical recycling, CO_2, and biomass are converted into the 14 plastics by a multitude of technologies. Technologies covered, for instance, comprise conventional fossil-based production technologies like steam cracking of hydrocarbon feedstock, polymerization of olefins and BTX, but also the CO_2-based methanol production or the gasification of biomass for synthesis gas production. Each technology is represented by a full life cycle assessment compliant dataset that includes the full mass and energy balances of the respective technology (cf. Section 5.3). Together with the supply of carbon feedstock, the production of chemicals and plastics is based on approx. 200 harmonized technology datasets that are compliant with the life cycle assessment standard.

Module 5 - end-of-life. Three options exist to treat plastic waste: landfilling, incineration (i.e., energy recovery), and recycling. Several intermediate waste fractions occur during these three waste treatment options, such as sorted fractions, residues from sorting, and residues from mechanical and chemical recycling (cf. Figure 5.3). Landfill and incineration are included for all plastic wastes and intermediate waste fractions. In the case of mechanical recycling, current practice from European countries like Germany and Austria shows that plastic packaging waste can be efficiently recycled mechanically (Consultic, 2015; Prognos AG, 2008). In contrast, other plastic wastes suffer significant difficulties due to impurities in the waste and, thus, cannot be adequately recycled mechanically (Ragaert et al., 2017). Thus, this thesis only considers mechanical recycling for sorted fractions of plastic packaging waste. In addition to mechanical recycling, chemical recycling is highlighted as one of the most promising technologies to combat the plastic crisis and, at the same time, reduce greenhouse gas emissions of chemical and plastic production (Rahimi and García, 2017; Clark et al., 2016; Nat Sustain, 2018). While chemical recycling is not limited with regard to the type of plastic waste input, as long as it has been cleaned from metal residues and ashes, its outputs range from refinery feedstock over liquid and gaseous fuels to monomer products and even value-added chemicals (Hong and Chen, 2017; Zhang et al., 2020). However, these technologies are still under development. Thus, the bottom-up model includes the pyrolysis of plastic wastes to steam-cracking feedstock, e.g., naphtha, and the production of the respective chemical monomers from chemical recycling as early-stage technologies. Chemical recycling is assumed to be able to treat every plastic waste and all intermediate waste fractions.

Final demand: The final demand defines the total output of the desired plastic that needs to be provided by the plastic production. Additionally, the final demand defines the amount of plastic waste that has to be treated by the end-of-life technologies. This final demand is equal for all pathways and calculated results from the bottom-up model. Thereby, the final demand ensures a fair comparison between all results since, in all cases, the same amount of plastics are produced and the same amount of plastic wastes are treated. In terms of the ISO-standard 14040 and 14044 (ISO 14040, 2021; ISO 14044, 2021), the final demands equal the functional unit of the comparisons conducted throughout this thesis.

Allocation principle: During chemical and plastic production, several production processes produce by-products that are not necessarily utilized in subsequent processes to produce plastics. One example is the stoichiometric production of caustic soda as a by-product of chlorine used for polyvinyl chloride. To solve this problem, this thesis follows the system expansion approach (see also Section 5.2 for the mathematical approach). Thus, all production systems, circular or linear, are required to produce the same amount of plastics and by-products. The reasoning behind including by-products is that they are also demanded by other markets than plastics. However, in some cases, a cut-off is assumed. This cut-off means that small amounts of by-products, below 1% of the overall production volume of plastics, are neglected in the final demand for all pathways. One example of such a cut-off is hydrogen cyanide that is produced as a by-product during acrylonitrile production.

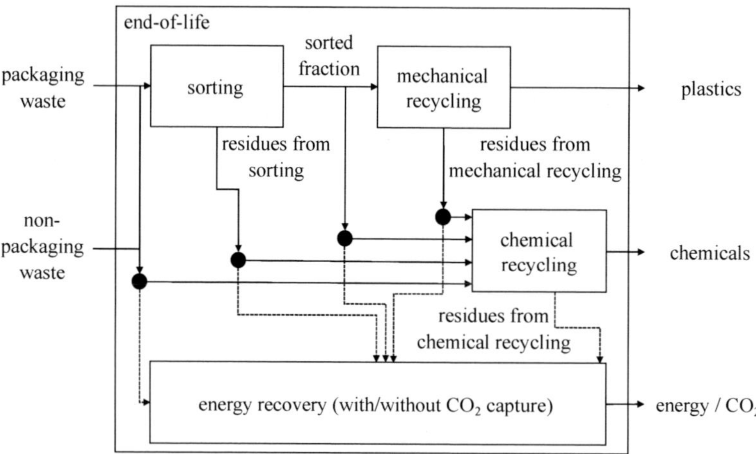

Figure 5.3: Overview of intermediate wastes during the end-of-life phase: sorted fractions, residues from sorting, and residues from mechanical and chemical recycling.

5.1.3 Using the bottom-up model beyond plastics

The focus of this thesis is on plastic production. However, the bottom-up model can be easily applied to other scopes and used to answer other scientific questions. Two examples are shown in Appendix D and E.

Appendix D focuses on the chemical industry and the potential of carbon capture and utilization technologies to reduce global greenhouse gas emissions of chemicals. Thus, Appendix D highlights the model flexibility to focus also on only one circular technologies and chemicals instead of plastics.

Appendix E focuses only on oxygenated products, like ethylene glycol or phenol, and the potential for novel electrochemical routes to reduce greenhouse gas emissions. In particular, the model is used to evaluate the most promising pathways for electrochemical processes that are in early-development or conceptual phases. Thus, Appendix E highlights that the model can also be used to identify the most promising pathways in the early stages of process development.

The two examples highlight that the model can be utilized to analyze different parts of chemical and plastic supply chains and, thus, offers high flexibility to answer several scientific questions.

5.2 Computational structure of the bottom-up model

The TCM (Kätelhön et al., 2016) is based on the general calculus of life cycle assessment (Heijungs and Suh, 2002; Hauschild and Huijbregts, 2015) and the following five fundamental elements: technologies, intermediate flows, elementary flows, characterization factors, and the final demand. Each of these elements is defined below:

(1) Technologies are activities that transform input flows into output flows. Input flows and output flows are categorized as intermediate flows or elementary flows (see below).

(2) Intermediate flows are exchanges of energy, materials, or products between technologies. Example: a chemical raw material, such as ethylene, used for a polymer, e.g., polyethylene.

(3) Elementary flows are exchanges of energy or materials between technologies and the environment. Example are: crude oil input from ground or greenhouse gas emissions to air.

(4) Characterization factors relate and transfer elementary flows into their impact on a particular environmental impact category. Example: The global warming potential of methane equals 28 kg CO_2-eq per kg methane (see table 8.A.1 in IPCC (2014)). Thereby, several greenhouse gas emissions can be related to the indicator kg CO_2-eq.

(5) The final demand is the total output or the amount of waste treatment of a set of intermediate flows. Example: 1 kg of polyethylene or 1 kg of waste polyethylene.

The general life cycle assessment matrix calculus (Heijungs and Suh, 2002) provides the structure for the data used in the bottom-up model. This data is structured in matrix A, and matrix B. Matrix A represents how technologies transform intermediate flows, while matrix B represents how technologies consume and emit elementary flows. In matrix A, technologies are represented by columns. Rows represent intermediate flows. The coefficient a_{ij} represents the intermediate flow i, that is related to a technology j. This relationship is either production (for $a_{\mathrm{ij}} > 0$) or consumption (for $a_{\mathrm{ij}} < 0$). The matrix A includes all datasets required to represent the full system boundary described in Section 5.1.2. Similarly, in matrix B the coefficient b_{ej} shows the elementary flow e in relationship with the technology j. This relationship will be either consumption (for $b_{\mathrm{ej}} < 0$) or emission (for $b_{\mathrm{ej}} > 0$). The final demand is considered using the vector y, where an element y_{i} represents the final demand for intermediate flow i, e.g., plastics and plastic wastes. For each intermediate flow, multiple production technologies are considered, allowing for a choice between technologies. The objective can be any environmental impact or cost across the entire related technology system, producing the given final demand y. For this purpose, a scaling vector s is defined that adjusts the quantities of intermediate flows in A in measure with the final demand.

This scaling vector s allows for the determination of the cumulative elementary flows g resulting from the production of the final demand, where

$$g = Bs. \tag{5.1}$$

In alignment with the guidelines for life cycle impact assessment, these accumulated elementary flows g are translated into environmental impact categories (ILCD, 2010a,b). The matrix Q represents these categories and includes the characterization factors. The element q_{ek} of Q represents the characterization factor of elementary flow e for the environmental impact category k. Thereby, the environmental impacts h for all impact categories can be calculated as:

$$h = QBs. \tag{5.2}$$

The optimization problem to solve the bottom-up model relies on these life cycle assessment definitions. In accordance with the scope of this thesis to minimize greenhouse gas emissions, the optimization problem is defined by the following equations:

$$\min h_{\mathrm{CO_2}} = Q_{\mathrm{CO_2}} B s \tag{5.3}$$
$$\text{s.t. } As = y_{\mathrm{t}} \tag{5.4}$$
$$0 \leq s_j \leq c \tag{5.5}$$

where $h_{\mathrm{CO_2}}$, represents the accumulated greenhouse gas emissions (in kg CO_2-equivalents), A is the technology matrix, y the final demand in year t, and c the potential upper bound for s. Equation 5.3 defines the objective to minimize the greenhouse gas emissions, while Equation 5.4 specifies the final demand being fulfilled. The objective function represents the accumulated equivalent CO_2 emissions. The elements in Q represent the 100-year global warming potential of each elementary flow, according to the IPCC (2014). The IPCC methodology allows for determining the equivalent greenhouse gas emissions of any given elementary flow relative to CO_2. Equation 5.5 adds the constraints that each scaling vector entry must be between zero and the upper bound c. The upper bound c can be used to, e.g., limit the supply of CO_2, biomass, or waste to a specific technology. Note that any other environmental impacts could be used as objective since the underlying datasets are fully life cycle assessment compliant.

An element i of the final demand y is defined as the total amount of a given intermediate flow, e.g., polypropylene, available for final consumption in the use-phase. Thus, the final demand represents the production volume of this intermediate flow minus the volume consumed by other production processes. However, often, only production volumes are available in the literature.

To calculate the final demand, a matrix L^{+}_{LVP} is used to specify the production outputs of intermediate flows from technologies included in the bottom-up model. Rows of L^{+}_{LVP} indicate large volume plastics. Columns illustrate each of the technologies included in the bottom-up model. Production inputs are set to zero in L^{+}_{LVP}. The element l_{mj} represents the amount of large volume plastic m produced by process j. The vector p_{t} is used such that only the respective production volumes of the 14 large-volume plastics for the specific year t is produced. Given L^{+}_{LVP} and p_{t}, the scaling vector s* can be calculated by solving the following optimization problem:

$$\min h_{\mathrm{CO_2}} = Q_{\mathrm{CO_2}} B s^{*} \tag{5.6}$$
$$\text{s.t. } As^{*} \geq 0 \tag{5.7}$$
$$L^{+}_{\mathrm{LVP}} s^{*} = p_{\mathrm{t}} \tag{5.8}$$

$$0 \leq s_j^* \leq c \tag{5.9}$$

The purpose of scaling vector s* is twofold: First, it scales the production volume of each production process such that the production volume of each plastic included in the bottom-up model is equal to the production volume in year t (Equation 5.8). Second, the same scaling ensures that the supply chain produces all inputs (Equation 5.7) in exactly the amount needed for the respective plastic production volumes.

Finally, the matrix $L_{\mathrm{LVP}}$is defined. Here, the columns denote each technology within the model, and the rows represent the 14 large volume plastics. An element l_{mj} of L_{LVP} represents the large volume plastic m produced or consumed by process j. y^{t} represents the final demand of the plastics included in the model in year t:

$$y_{\mathrm{t}} = L_{\mathrm{LVP}} s^*. \tag{5.10}$$

5.3 Data to build the bottom-up model

This section provides a more detailed explanation of the data used in the bottom-up model. For this purpose, this thesis follows the structure of the optimization model from Section 5.2. First, the data to represent the technologies, e.g., matrices A and B, are described in Section 5.3.1. The data is presented for each module inside the system boundary (cf. Section 5.1.2) separately. Second, Section 5.3.2 explains the calculation of the production volumes of plastics and plastics wastes (cf. p_{t} in equation 5.8) and that are needed to calculate the final demand (cf. y in year t in equations 5.4 and 5.10). A complete list of technologies, datasets, and their mass and energy balances that can be made publicly available, is available on Zenodo (Raoul Meys, 2021).

5.3.1 Technology datasets

Technologies transform input flows into output flows. Both input and output flows can be intermediate flows (included in A) or elementary flows (included in B). In addition to the distinction between intermediate and elementary flows, the datasets for technologies are either aggregated or unit datasets.

(1) Aggregated datasets: A fully terminated aggregated dataset is a dataset that represents the entirety of a product system. In aggregated datasets, only elementary flows enter or leave the system, and product flows leave. Any other technical flows are created and consumed within the production system itself and

therefore do not feature in the terminated aggregated dataset. In the bottom-up model, most aggregated datasets represent the global production mix (global aggregated dataset). Where a global aggregated dataset is not available, an European aggregated dataset is used in its place.

(2) Unit datasets. Unlike aggregated datasets, unit datasets represent the smallest possible part of a process for which input and output data can be defined. Unit datasets only consider the intermediate and elementary flows of a single unit process. As data is not aggregated, the full mass and energy balance are depicted in the dataset.

In the following, the method used to generate the datasets used in each module of the system boundary are briefly described.

Module 1 - CO_2 supply. The CO_2 supply is based on unit datasets that have been derived based on literature research. In the case of carbon dioxide supply from ammonia, ethylene oxide, and hydrogen producing plants, as well as biomass fermentation and gasification, the respective unit processes are used.

For ammonia and hydrogen-producing plants, the Rectisol process is used to capture CO_2 since it is one of the major applied processes on an industrial scale (IHS Markit, 2018). The Rectisol is also applied for the biomass gasification processes (Arvidsson et al., 2014; Hannula and Kurkela, 2010, 2012; Isaksson et al., 2012; Spath et al., 2005). This Rectisol process utilizes cryogenic methanol at a minimum of minus 40°C in which CO_2 and hydrogen sulfides are dissolved and subsequently recovered in two separation columns.

In the case of biomass fermentation, almost pure CO_2 is generated (Humbird et al., 2011) that must be freed from some minor impurities and water via molecular sieve adsorbents.

For waste incineration, CO_2 is captured by the monoethanolamine process (Haaf et al., 2020). Here, aqueous monoethanolamine is circulated between an absorber and a desorber. In the absorber, CO_2 is absorbed into the aqueous monoethanolamine solution via a chemical reaction. Subsequently, the mixture is heated such that CO_2 is dissolved in the desorber section.

In the case of ethylene oxidation with oxygen to produce ethylene oxide, CO_2 is captured via the potassium carbonate process. The potassium carbonate process is based on a reversible reaction of potassium carbonate, water, and carbon dioxide to potassium bicarbonate. In an absorber, potassium bicarbonate is formed at lower temperatures. The resulting bicarbonate solution is then fed into the top of a regenerator in which the temperature is increased such that the CO_2 is released from the solution, and potassium carbonate is formed again (Elvers and Ullmann, 2011; IHS

Markit, 2018).
The data used to model the direct air capture process is based on the Climeworks plants in Hellisheioi and Hinwil. This process is based on an adsorption-desorption cyclic principle. In the adsorption phase, CO_2 reacts chemically with the adsorbent. During the following desorption phase, the adsorbent is heated, and a vacuum is produced. As a result, the CO_2 is removed from the adsorbent. Water from the CO_2 stream is separated by cooling. Here, it is assumed that the current direct air capture system uses heat from heat pumps and is supplied with wind-based electricity. Thus, per kg of captured CO_2, 4.7 MJ of thermal energy and 0.7 kWh must be supplied to the direct air capture system (Deutz and Bardow, 2021).
In all cases, the CO_2 needs to be pressurized from ambient conditions to 100 bar. Data for the pressurization was taken from Farla et al. (1995).

Module 2 - biomass supply. Regarding the biomass supply, this thesis uses aggregated datasets from the standard life cycle assessment database ecoinvent 3.6 (Ecoinvent, 2020) that includes all elementary flows for harvesting and producing the respective lignocellulose feedstock. However, the elementary flows do not include the carbon uptake of the biomass. Thus, the uptake of CO_2 (cf. Table 5.1) has been added to the datasets. The ecoinvent data considers emissions from land-use-change, and the analysis assumes that absolute indirect land-use-change greenhouse gas emissions for biomass stay the same as today. This assumption is in line with the recent evaluation of global strategies to mitigate greenhouse gas emissions of plastic life cycles (Zheng and Suh, 2019). Furthermore, it is assumed that biomass is made available from marginal land and, thus, no land elsewhere must be repurposed. Thus, changes in greenhouse gas emissions due to indirect land-use are excluded. However, future assessments should carefully evaluate the use of lignocellulose feedstock and the respective influence of indirect land-use change effects on the greenhouse gas emissions of bio-based plastic production.

Module 3 - fossil resource supply. The elementary flows of fossil resource supply are included in the bottom-up model: from raw materials extraction, e.g., crude oil or natural gas, over distribution to downstream processing (Althaus et al., 2007). The datasets are based on aggregated datasets from the ecoinvent 3.6 database (Ecoinvent, 2020).

Module 4 - chemical and plastic production. Since over 200 datasets are used to represent chemical and plastic production, only the methodology of dataset selection and a brief overview of major literature sources is given here. The selection of datasets is based on a step-wise procedure illustrated in Figure 5.4.

In step 1, the conventional best available technologies are identified that produce at

Table 5.1: Summary of the composition of biomass resources and their CO_2 uptake per kg of biomass.

property	wood chips	wood pellets	bark chips	miscanthus
Moisture [wt.%]	50	8	50	14
Ash [wt.%]	2.2	0.3	3.2	2.9
Ultimate analysis, dry basis [wt.%]				
C	50.3	50.7	52.6	48.3
H	5.4	6.1	5.1	6
N	0.5	0	0.4	0.5
Cl	0	0	0	0.2
S	0	0	0.4	0.1
O	41.6	43.9	38.2	42
CO_2 uptake [kg CO_2]	1.84	1.86	1.93	1.77

least one of the included chemicals or plastics (cf. Section 5.1.1). These best available technologies are identified based on publicly available datasets based on the lowest greenhouse gas emissions (Ausfelder et al., 2013) and the European Commission's reference documents (European Commission. Joint Research Centre., 2007, 2017). By including only the best available technologies, the calculated greenhouse gas emissions represent a lower bound for conventional and fossil-based plastic productions since they lead to minimal greenhouse gas emissions. Exceptions are reported in Appendix C.

In step 2, each best available technology's mass and energy balances are included as a unit dataset (2a) if the respective technology consumes at least one of the included chemicals and plastics. If the best available technology does not consume any of the chemicals and plastics, an aggregated dataset is included (2b). Unit datasets for the conventional production technologies were derived from the process database IHS PEP Yearbook (IHS Markit, 2018), except for the production of polyurethanes (Ecoinvent, 2020) and steam cracking processes (CarbonMinds, 2020). In the IHS PEP Yearbook, all datasets have been verified by industrial experts and, thus, offer a very sophisticated source of data based on industrially validated process simulations. However, the datasets do not include elementary flows and waste flows, such as inorganic or organic residues.
Thus, an existing waste incineration model (Doka, 2003, 2013) has been modified to represent state-of-the-art waste treatment in chemical parks. The resulting incineration model represents hazardous waste incineration by integrating industrial data from

hazardous waste incineration plants in a chemical park in western Germany (Chemical park operator, 2018). In that chemical park, wet scrubbers, electrostatic precipitators, and DeNOx stages (SCR-low dust and SNCR) are used for flue-gas cleaning. Based on actual measurements, the respective transfer coefficients of the waste incineration model were updated to represent the industrial standard in chemical parks. A more detailed description of the energy recovery and waste incineration model can be found in Appendix A.

In step 3, biomass- and CO_2-based production technologies for the chemical and plastic scope are identified based on studies by the International Energy Agency (Bazzanella and Ausfelder, 2017) and the European Chemical Industry Council (Ausfelder et al., 2013). All datasets representing the biomass- and CO_2-based production technologies are included as unit datasets. The several scientific literature data sources are summarized in Appendix C and the technologies data that is not proprietary is presented on Zenodo (Raoul Meys, 2021). The methodology to close data gaps for early-stage technologies is provided in Section 5.4.

In step 4, all missing inputs of the unit datasets are determined. Additionally, aggregated datasets for missing inputs are included to enable a fully consistent supply chain of chemicals and plastics.

In step 5, energy recovery technologies of high-calorific production wastes are included inside the bottom-up model. These energy recovery technologies are based on the same waste incineration model as described in step 2. Additionally, waste incineration to estimate the environmental impacts of disposing unneeded by-products without energy recovery (e.g., flaring) is included.

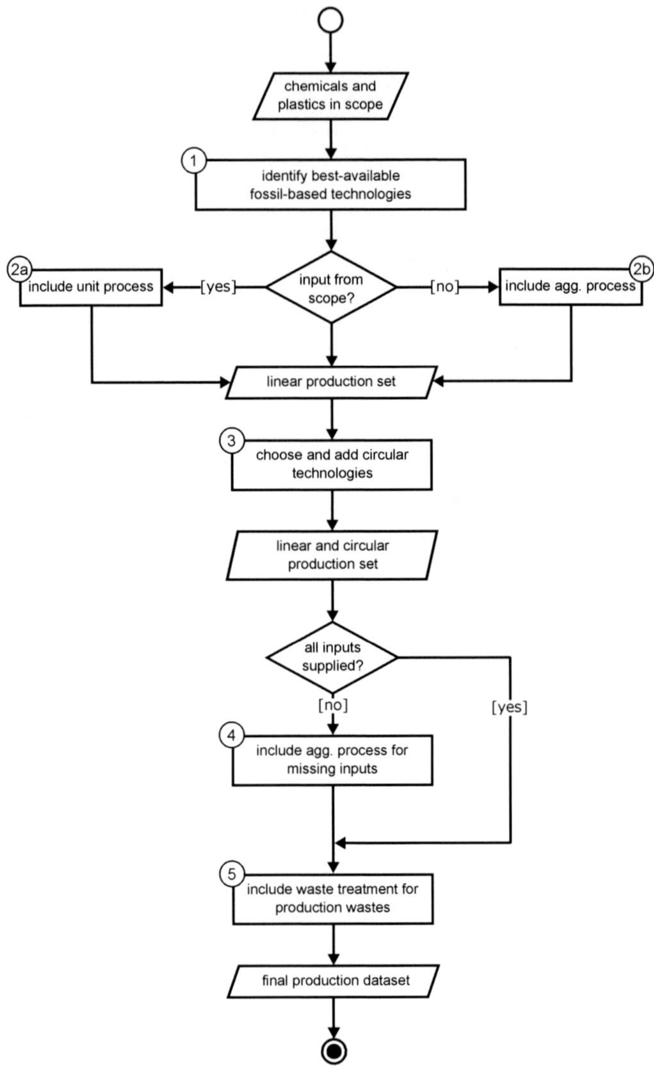

Figure 5.4: Collection procedure for chemical and plastic production datasets.

Module 5 - end-of-life. In the end-of-life module, unit datasets are included for waste incineration as well as chemical and mechanical recycling. For waste incineration, Doka's energy recovery model (Doka, 2003, 2013) forms the basis for modeling unit datasets. To model the mechanical recycling of plastic packaging waste, industry data from 2017 (HTP GmbH & Co. KG, 2017; Dehoust et al., 2016) is used. In the case of chemical recycling, datasets representing pyrolysis from plastic waste to liquid hydrocarbons as steam cracker feedstock are based on industrially validated process simulations (IHS Markit, 2018). In the case of monomer production from plastic waste, a simplified approach to generate unit process datasets according to Section 5.4 is used. In contrast to waste incineration and recycling, landfill of plastic wastes is based on aggregated datasets (Ecoinvent, 2020).
A more detailed description of mechanical and chemical recycling, as well as waste incineration can be found in Appendix A and Appendix C.

5.3.2 Production volumes of plastics and plastic wastes

The calculation of production volumes of plastics is based on the initial values derived by Geyer et al. (2017) for the year 2015. However, in some cases, no specific production volumes could be obtained directly. Thus, in the case of polystyrene, polyurethane, polyethylene, and the fiber plastics polyethylene terephthalate, polyamides, and polyacrylonitrile, market shares reported by Zheng and Suh (2019) are used. In the case of polyamides, production volumes are further subdivided into polyamide 6 and 66 by using their respective shares from the year 2016 (PlasticsInsights, 2016). To calculate future production volumes of plastics, this thesis uses data of ICIS (2018) from the year 2000 (PV_{2000}) and a forecast for 2040 (PV_{2040}) since no data for 2050 was available. Using Equation 5.11, the annual growth rates GR_{LVP} for each of the 14 plastics are calculated based on PV_{2000} and PV_{2040}:

$$GR_{\text{LVP}} = (\frac{PV_{2000}}{PV_{2040}})^{(1/40)}. \qquad (5.11)$$

Table 5.2 shows the production volume for each plastic in 2015 and 2050, as well as the annual growth rates (ICIS, 2018).

In this thesis, the life cycle accumulated greenhouse gas emissions of 90% of the global plastics are calculated. Thus, the volume of plastic waste to be managed is equal to the respective plastic production in 2050. In reality, however, the amount of plastic waste each year is lower than the amount of plastic production because, on average, plastics do not reach the end of life in the same year that they are produced.

Table 5.2: Table of production volumes in 2015 and 2050, and annual growth rates for the 14 large-volume plastics (Geyer et al., 2017; Zheng and Suh, 2019; ICIS, 2018).

plastics	production volume in Mt per year		Annual growth rate in %
	2015	2050	
polyamide 6	2.7	5.1	1.9
polyamide 66	2.1	10.6	4.8
PET pellets (fiber-grade)	85.3	327.5	3.9
PET pellets (bottle-grade)	33	170.5	4.8
polyacrylonitrile fiber	2.1	4.7	2.4
polyethylene, HD	52	174.7	3.5
polyethylene, LD	27.5	45.9	1.5
polyethylene, LLD	36.5	181.1	4.7
polypropylene	68	285	4.2
polystyrene, GP	12.5	19.1	1.2
polystyrene, HI	12.5	19.1	1.2
polyurethane, flexible	15	33.9	2.4
polyurethane, rigid	12	27.2	2.4
polyvinyl chloride	38	113	3.2

Instead, they are used for different purposes and therefore have different life spans. An illustrative example is a comparison of plastic packaging and plastics in the building environment. While plastics in packaging have an average lifespan of half a year, plastics used in construction have an average lifespan of 35 years (Geyer et al., 2017) and, thus, enter the end-of-life at a later point of time. If the actual plastic waste volumes in 2050 were taken into account, the accumulated greenhouse gas emissions of plastic production and waste treatment in 2050 based on captured CO_2 and biomass would be negative since more carbon would be incorporated into the production than emitted from the incineration of plastics. Since net-negative greenhouse gas emissions of products can lead to false conclusions about the benefits of CO_2 or biomass utilization, this thesis focuses on the life cycle greenhouse gas emissions of plastics. Note that net-negative greenhouse gas emissions over the life cycle of plastics by CO_2 or biomass utilization can only be achieved by permanent carbon storage (Gabrielli et al., 2020).

5.4 Closing data gaps for missing technology datasets

For some chemical recycling and some CO_2 utilization technologies, unit datasets are unavailable because these technologies are in early development stages. Therefore, unit process data based on a stoichiometric reaction are determined, which is standard practice in life cycle assessment studies where data gaps exist (Sugiyama et al., 2008; Patel et al., 2012; Geisler et al., 2004). Calculating unit process data follows the stoichiometric reaction:

$$v_{\mathrm{i}_1} M_{\mathrm{i}_1} + ... + v_{\mathrm{i}_\mathrm{n}} M_{\mathrm{i}_\mathrm{n}} \rightarrow v_{\mathrm{p}_1} M_{\mathrm{p}_1} + ... + v_{\mathrm{p}_\mathrm{m}} M_{\mathrm{p}_\mathrm{m}}. \tag{5.12}$$

Here, v represents the stoichiometric coefficient, indices $\mathrm{i}_{1,...,\mathrm{n}}$ reactants, indices $\mathrm{p}_{1,...,\mathrm{m}}$ products, and M the molar mass. With reaction 5.12 as a foundation, this thesis then determines the mass flow m_i of each reactant i:

$$m_\mathrm{i} = \frac{v_\mathrm{i} M_\mathrm{i}}{v_\mathrm{p} M_\mathrm{p}}. \tag{5.13}$$

To account for inefficiencies such as incomplete reactions or loss of product, this thesis follows the guidance of Hischier et al. (2005) and assumes a product yield X of 95 wt.%. The difference of 5 wt.% is assumed to be sent to industrial waste treatment, e.g., incineration. In this way, the mass flow $m_{\mathrm{i}_\mathrm{p}}$ of each input can be represented using the following equation:

$$m_{\mathrm{i}_\mathrm{p}} = m_\mathrm{i} \frac{1}{X}. \tag{5.14}$$

Furthermore, as chemical processes require energy for operation and purification, the energy demand must be included. To include energy demands, this thesis uses a simplified approach. First, the missing energy inputs are estimated based on the average energy demand for chemicals produced at Gendorf, a German chemistry park, where more than 30 companies produce approximately 1500 different chemicals Althaus et al. (2007). Accordingly, for thermal energy demand Q_H and electricity demand P_el, average demands of 2 GJ and 1.2 GJ are used per ton of product. Secondly, standard thermodynamic relations to calculate the minimum energy requirement $Q_\mathrm{H,min}$ as detailed in equation 5.15 are used. Following Hess' law of constant enthalpy summation and assuming that water is in a liquid state, the energy balance calculates the minimum energy requirement $Q_\mathrm{H,min}$ based on the gross calorific values of products $\Delta h^0_\mathrm{c,j}$ and reactants $\Delta h^0_\mathrm{c,i}$:

$$Q_\mathrm{H,min} = \sum\nolimits_\mathrm{j} m_\mathrm{j} \Delta h^0_\mathrm{c,j} - \sum\nolimits_\mathrm{i} m_\mathrm{i} \Delta\Delta h^0_\mathrm{c,i}. \tag{5.15}$$

As a final step, this thesis checks if the sum of energy inputs Q_H and electricity demand P_el is lower than the minimal energy requirement $Q_\mathrm{H,min}$. If it is the case, the difference of $Q_\mathrm{H,min}$ and the sum of energy inputs Q_H and electricity demand P_el

are added to the thermal energy demand of the respective process. By this means, it is assured that no unit process breaks the energy balance 5.15 by using the average Gendorf values.

Chapter 6

The fossil and linear plastic life cycle

This chapter presents and analyzes the results calculated for the fossil and linear plastic life cycle based on the bottom-up model described in Chapter 5. To do so, Section 6.1 shows the greenhouse gas emissions and energy consumption for the year 2015, based on current rates of waste treatment technologies. The results based on the bottom-up model are compared with literature depicting the plastic supply chain, energy demands (Ausfelder et al., 2013; Bazzanella and Ausfelder, 2017; IEA, 2018a) and recent estimates by Zheng and Suh (2019) about the global greenhouse gas emissions of plastics. By this means, the validity of the model is evaluated against the current status of literature. Afterward, Section 6.2 highlights the future greenhouse gas emissions and energy demands of the plastic supply chain for the year 2050. These energy demands and greenhouse gas emissions in 2050 serve as the benchmark for the subsequent Chapter 7 in which renewable and circular plastics are assessed.

6.1 Energy consumption and greenhouse gas emissions in 2015

By solving the optimization model described in Section 5.2, the complete supply chain of the fossil-based and linear plastics production is generated by the bottom-up model for the year 2015 (cf. Figure 6.1). This supply chain starts with the production of ethylene, propylene, benzene, toluene and xylene (denoted as high-volume chemicals) from fossil resources. In total, 324 Mt of high-volume chemicals are produced via the steam cracking of naphtha (239 Mt) and the solvent extraction plus crystallization of p-xylene (85 Mt). These values are in line with reported global production volumes of high-volume chemicals by the International Energy Agency and CEFIC (Ausfelder et al., 2013; Bazzanella and Ausfelder, 2017; IEA, 2018a), summing up to 308 Mt of global production. In order to evaluate how good the overall mass flows of the bottom-up model correlate with the actual mass flow of the current chemical and plastic industry, we calculate the absolute difference between the model results and the values derived in Section 2.2.1. While this number does not represent the greenhouse gas emissions, it at least can be used to confirm correspondence of the energy flows of the model with the global chemical and plastic supply chain. For ethylene and propylene, the bottom-up model leads to a direct conversion of 116 Mt to polyethylene (73%) or 68 Mt to polypropylene (85%). The percentages are only 3 or 5% higher than reported values by ICIS (2018) for the year 2018 (cf. Table 6.1)[1]. Considering all utilization ratios of all high-volume chemicals, the calculated values based on the bottom-up model lead to an absolute average difference of 4% over all high-volume chemicals compared to the values from ICIS (2018). The difference is most probably based on the fact that the bottom-up model does not include all polymer products. This assumption is further confirmed by the lower utilization shares of all high-volume chemicals for other polymer products. Utilization rates in other polymer products are underestimated between 5 and 11% for all high-volume chemicals. This lack of detail to depict smaller-scale plastic products could potentially be overcome by including more than 14 plastics in the bottom-up model. However, the effort to depict even smaller-scale plastic applications is relatively high due to their very complex supply chains. Since the absolute average difference of utilization ratios is only 4%, focusing on the 14 plastics seems a reasonable and good compromise between model accuracy and modeling effort.

Overall, the fossil-based production of plastic materials requires 1 EJ of electricity and 25.2 EJ oil-eq of fossil resources. Assuming the current global average value of fossil resource depletion for electricity (2.38 EJ oil-eq per EJ electricity), the total equivalent fossil energy consumption equals 27.58 EJ oil-eq. The International Energy Agency reports that the global chemical industry used 31.3 EJ oil-eq in 2015 (IEA, 2018a). Assuming the average utilization of 84% of chemical products in polymers, the global energy consumption estimated by the International Energy Agency corresponds to 26.29 EJ oil-eq for polymers in 2015. This value is only 5% lower than the overall energy consumption calculated based on the bottom-up model in this thesis. While the overall energy consumption calculated with the bottom-up model seems to match recent estimates by the IEA, the type of fossil feedstock differs slightly: Currently, approx. 4% or 1% of the high-volume chemical production is based on either coal or natural gas, respectively (IEA, 2018a). The calculated shares of coal and natural gas as chemical feedstock based on the bottom-up model of this thesis are below 0.1%.

[1]Unfortunately, no values have been obtained for 2015 from ICIS. However, it is assumed that the difference between 2015 and 2018 for the utilization shares are relatively small.

Thus, similarly to the utilization ratios of high-volume chemicals, the fossil resource supply also slightly differs from current reported values. The major difference is the lack of ethylene and propylene production from methanol that is current practice in China. Since only the best available production technologies have been included in the bottom-up model (cf. Section 5.3), coal-based methanol has been excluded from this thesis. Furthermore, the production of natural gas-based methanol and its subsequent usage to produce ethylene and propylene leads to higher greenhouse gas emissions than the conventional steam cracking processes and, thus, is not chosen by the optimization algorithm. This small shift from methanol (in China, largely coal-based) to oil-based ethylene and propylene production via steam cracking results in a slight variation of the current fossil-resource shares used in the chemical and plastics industry.

Table 6.1: Summary of utilization ratios of high-volume chemicals for plastic products for ICIS (2018) and this thesis in wt.%. The value in brackets for ICIS (2018) equals the utilization ratios for all production volumes, i.e., including the production of other chemical products (cf. also Section 2.2). BTX include benzene, toluene and xylene.

chemical	application	ICIS (2018)	this thesis	difference
ethylene	polyethylene	70% (61%)	73%	3%
ethylene	other polymers	6% (5%)	1%	-5%
ethylene	polyvinyl chloride	10% (9%)	11%	1%
ethylene	polyesters	10% (9%)	10%	0%
ethylene	polystyrene	3% (3%)	4%	1%
propylene	polypropylene	80% (66%)	85%	5%
propylene	other polymers	13% (10%)	6%	-7%
propylene	polyurethanes	7% (6%)	9%	2%
BTX	polyesters	54% (43%)	59%	5%
BTX	polystyrene	23% (19%)	23%	0%
BTX	other polymers	16% (13%)	5%	-11%
BTX	polyurethanes	7% (6%)	13%	6%
		average absolute difference		4%

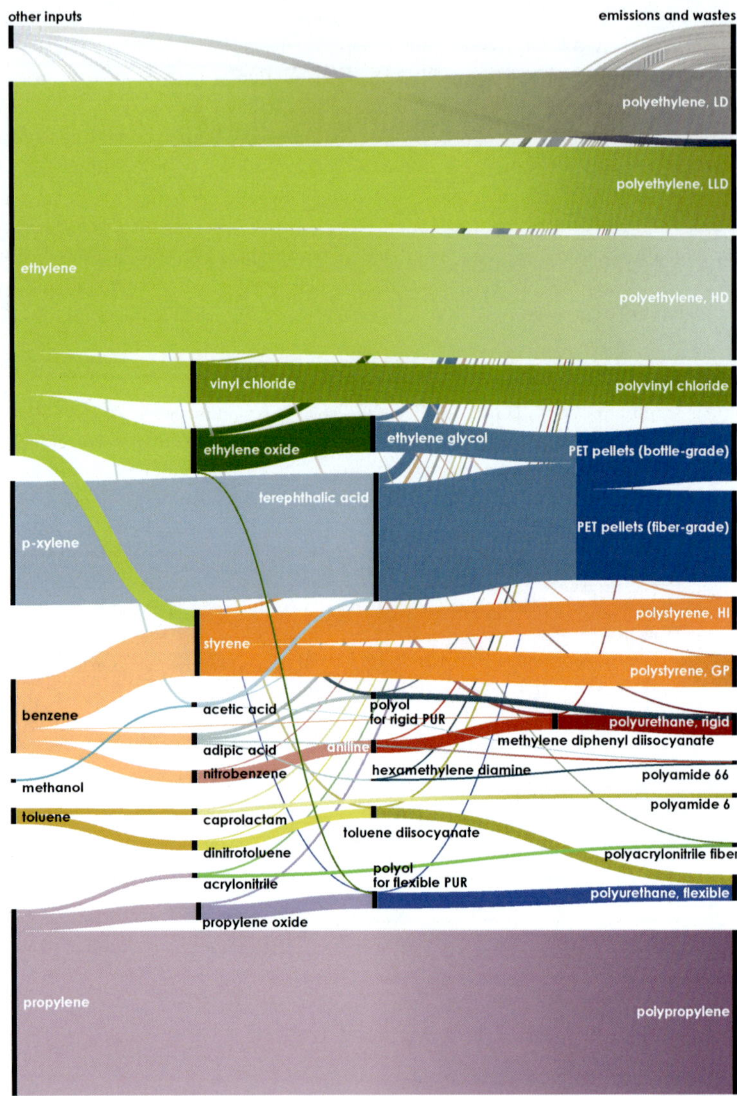

Figure 6.1: The supply chain of plastic production is shown for the year 2015 and the linear-carbon pathway. The width of each line represents the relative carbon content of each chemical flow. Only the major chemical intermediates are shown to increase readability.

Based on the extensive use of fossil resources, plastic production would lead to 826 Mt of CO_2-eq greenhouse gas emissions (cf. Figure 6.2 left bar) based on the bottom-up model. These greenhouse gas emissions assume 0.15 kg of CO_2-eq greenhouse gas emissions per MJ of supplied electricity, which is equal to the current global average carbon grid intensity (Ecoinvent, 2020). Furthermore, assuming current shares of waste incineration (24%), recycling (20%), and landfill (56%), the bottom-up model calculates 260 Mt of CO_2-eq of additional greenhouse gas emissions from plastic waste treatment. While landfill does not produce valuable products, waste incineration and recycling produce electricity and heat as well as recycled plastic resins. These valuable products have to be accounted for. Assuming a credit for electricity production (0.21 EJ, credit of 0.15 Mt of CO_2-eq per EJ (Ecoinvent, 2020)) and thermal energy (1.05 EJ, credit of 0.056 Mt of CO_2-eq per EJ (Ecoinvent, 2020)), the greenhouse gas emissions of waste treatment are reduced to 170 Mt of CO_2-eq greenhouse gas emissions. While this calculation does not consider any credit for recycled plastics, it is, however, consistent and comparable with recent estimates of the global greenhouse gas emission of plastic life cycles by Zheng and Suh (2019). Zheng and Suh (2019) calculate that production emissions of global plastics equal 1085 Mt CO_2-eq while end-of-life contributes with 0.16 Mt CO_2-eq (cf. Figure 6.2 right bar). Fixing the scope by Zheng and Suh (2019) to the scope of this thesis by excluding additives and other polymers, the comparable greenhouse gas emissions equal 969 and 146 Mt CO_2-eq for the production and end-of-life, respectively (cf. Figure 6.2 middle bar). Considering identical values for greenhouse gas emissions of converting polymer resins to plastic products for this thesis and Zheng and Suh (2019), the results obtained in this thesis differ by 8% (comparison of left and middle bar in Figure 6.2).

The difference can be explained by the optimization approach that minimizes greenhouse gas emissions. The values by Zheng and Suh (2019) rely on aggregated values from Ecoinvent (2020) that depict the current way of producing plastic materials. In contrast, the optimization approach minimizes greenhouse gas emissions based on the optimal combination of current best-available technologies. As a result, the overall greenhouse gas emissions are expected to be lower than the current plastic life cycle. However, by using the optimal combination and, thus, the minimal greenhouse gas emissions, the bottom-up model of this thesis leads to a lower bound, and thus, robust estimations of potential reductions of greenhouse gas emissions by novel circular technologies. Furthermore, in contrast to Zheng and Suh (2019), the detailed supply chain from fossil chemical feedstock over chemicals to plastics can be fully depicted by the novel bottom-up model. Finally, Zheng and Suh (2019) show large-scale reduction potentials by increasing recycling rates or utilizing bio-based plastics (up to 1990 levels), but do not consider carbon capture and utilization technologies and do not provide pathways towards net-zero emission plastics.

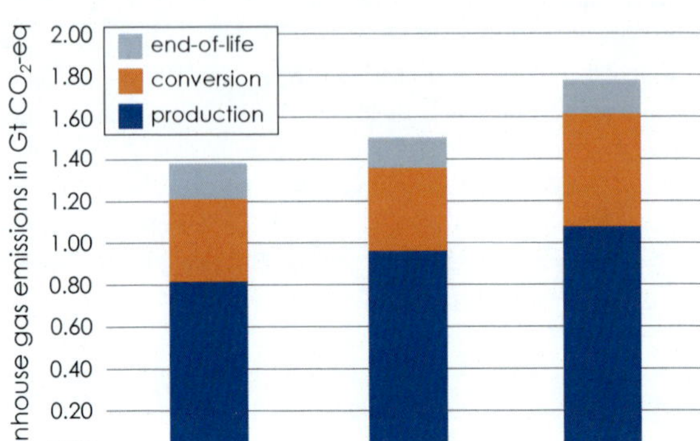

Figure 6.2: The greenhouse gas emissions for the plastic life cycle are shown for this thesis and Zheng and Suh (2019) for the adjusted scope and the original scope. The fixed scope of Zheng and Suh (2019) is comparable to the scope of this thesis. Greenhouse gas emissions for the conversion of polymer resins to plastic products are assumed equal for this thesis and the adjusted scope of Zheng and Suh (2019) in order to enable an equal comparison.

6.2 The linear-carbon pathway in 2050

After benchmarking the results obtained from the bottom-up model with current literature for 2015, the greenhouse gas emissions and energy demands are calculated for the future life cycle of plastics in 2050. These future greenhouse gas emissions and energy demands serve as a benchmark for the pathways applying circular technologies in the subsequent Chapter 7. The linear and fossil-based plastic life cycle is denoted as ***"linear-carbon pathway"*** and is based on the combination of fossil-based and conventional production technologies with waste incineration as a major waste treatment option. Thus, this thesis assumes that 94% of the plastic wastes are incinerated at their end-of-life. Only 6% are disposed in landfills, based on the minimal landfill rate reported by Geyer et al. (2017) for the year 2050. Even though current global recycling rates equal approx. 18 %, the linear-carbon pathway is based on incineration only in order to provide a theoretical upper bound for greenhouse gas emissions. The assumption of waste treatment shares, in particular the increase of recycling shares, is evaluated in the following Chapter 7.

By 2050, global plastic production equals 1417 Mt of plastic products (cf. Table 5.2 for more details). The bottom-up model leads to a complete supply chain from chemical feedstock over chemical to plastic production. While the quantitative amounts of chemical feedstock and chemicals change due to increased production volumes, the production technologies in the linear-carbon pathway are identical for the years 2015 and 2050. Thus, the production and utilization ratios are not described again.

Overall the linear-carbon pathway leads to an energy consumption of 3.96 EJ of electricity, 5.96 EJ of steam (equivalent to 1.73 Gt of steam) and 3.33 EJ of thermal energy to produce the 14 plastic materials (cf. Figure 6.3). Assuming an overall energy efficiency of 41% and a share of thermal energy of 74% for energy recovery from plastic waste (Eriksson and Finnveden, 2009), the incineration of the global plastic volume could produce 4.02 EJ of electricity and 12.38 EJ of thermal energy. Thus, the integration of waste incineration with plastic production could effectively substitute the process energy requirements for plastic production. In fact, 3.15 EJ of excess energy would be available to other sectors such as residential heating. If it is assumed that there is no integration between plastic production and energy recovery from plastic wastes, the overall energy demand of the production of plastics would equal 85.86 EJ oil-eq and 3.96 EJ of electricity. If the waste incineration and the production of plastics are integrated, the external fossil energy supply of the linear-carbon pathway equals 76.9 EJ oil-eq. Electricity is fully supplied via the incineration and energy recovery of plastic waste.

Based on the bottom-up model, the linear-carbon pathway (including energy integration from waste incineration), leads to 4.7 Gt CO_2-eq greenhouse gas emissions (cf. Figure 6.4). Specifically, 3.1 Gt CO_2-eq greenhouse gas emissions (66%) are caused by the waste incineration and 1.6 Gt CO_2-eq greenhouse gas emissions (34%) are caused by the production of plastic materials. Of these 34%, 31% are due to the production of chemical feedstock and chemical intermediates (e.g., monomers), while the conversion from monomers to plastics represents only represents 3%. The greenhouse gas emissions of chemical production are dominated by the production of chemical feedstock (13%) and direct emissions during fossil feedstock conversion to chemicals (18%), e.g., by steam cracker operations. Most of the contribution to greenhouse gas emissions of converting monomers to plastics is based the treatment of processing wastes (2%).

In order to highlight the importance of strategies for net-zero emission plastics, the bottom-up model results are compared to the global, total, and annual anthropogenic greenhouse gas emissions in 2050. Depending on the future scenario (Post-COVID-19

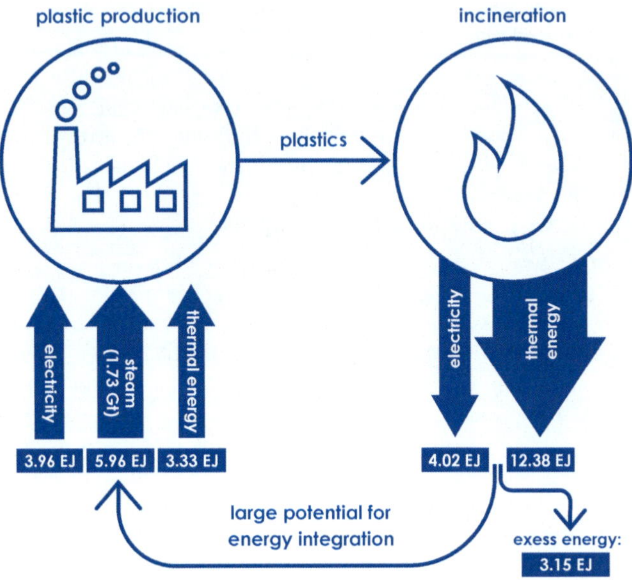

Figure 6.3: The energy consumption in terms of electricity, steam and thermal energy is shown for the production of plastic materials in the linear-carbon pathway. Furthermore, the production of thermal energy and electricity from the incineration of plastic wastes is shown.

and 1.5°C consistent), the annual global greenhouse gas emissions are predicted to be 59 to 7 Gt CO_2-eq emissions by 2050 (Roser and Ritchie, 2021). Thus, the plastic life cycle could contribute 9 to 67% of the yearly anthropogenic greenhouse gas emissions if no disruptive changes are implemented. To achieve these disruptive changes, circular technologies can be utilized. The contribution of these circular technologies to achieve net-zero emission plastics is assessed in the following Chapter 7.

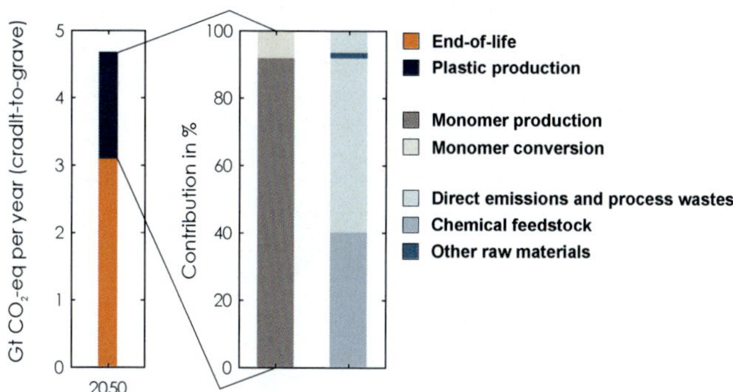

Figure 6.4: Life cycle greenhouse gas emissions of the linear-carbon pathway in 2050 and breakdown of greenhouse gas emission of plastic production into (1) monomer production and monomer conversion to plastics and (2) direct emissions and process wastes, chemical feedstock, and other raw materials. Other raw materials, for instance, include chlorine or ammonia.

Chapter 7

Achieving net-zero emission plastics by a circular carbon economy

Chapter 6 shows that reducing plastic's life cycle greenhouse gas emissions is crucial to achieve global climate targets. Strategies to mitigate greenhouse gas emissions from plastics include the decarbonization of energy supply and the implementation of circular technologies (Bazzanella and Ausfelder, 2017; Zimmerman et al., 2020) such as (1) chemical and mechanical recycling, (2) biomass utilization, and (3) carbon capture and utilization.

Recent literature on individual or partly combined circular technologies shows large-scale potential to reduce greenhouse gas emissions (Galán-Martín et al., 2021; Hermann et al., 2007; Posen et al., 2017; Schwarz et al., 2021; Zheng and Suh, 2019). However, no study identifies how circular technologies can be combined to achieve net-zero emission plastics. Furthermore, the current utilization of circular technologies is limited since circular technologies are generally associated with higher energy demands and costs (Gao et al., 2021).

This chapter shows that by combining recycling, biomass utilization, and carbon capture and utilization, net-zero greenhouse gas emission plastics can be achieved with lower energy demands and lower operational costs than the linear-carbon pathway (Chapter 6) combined with carbon capture and storage (CCS).

Based on the bottom-up model in Chapter 5, life cycle greenhouse gas emissions of plastic from cradle-to-grave for the year 2050 and five pathways are calculated (Section 7.1). The recycling pathway allows maximal recycling of all plastic wastes based on a minimal landfill rate of 6% reported by Geyer et al. (2017). In contrast, the biomass and the carbon capture and utilization (CCU) pathways assume that plastic waste is primarily incinerated. The resulting CO_2 emissions are circulated via biomass uptake or carbon capture and utilization. The circular-carbon pathway optimally combines recycling, biomass utilization, and carbon capture and utilization. The greenhouse gas emissions of all circular pathways are benchmarked with the linear-carbon pathway (Chapter 6). The current fossil-based industry includes recycling already today. Since landfilling of plastics will increasingly fade out (Geyer et al., 2017), the remaining options for plastic waste treatment are energy recovery (represented by the linear-carbon pathway) and recycling (represented by the recycling pathway). Thus, these two pathways depict the complete range of potential fossil-based futures for plastic waste treatment.

In Section 7.2, the energy demands for the circular-carbon pathway are benchmarked with the linear-carbon pathway in combination with large-scale carbon capture and storage. Afterward, Section 7.3 estimates the operational cost of both the circular-carbon and linear-carbon pathway. In the final Section 7.4, policy measures are discussed that can accelerate the transition to net-zero emission and circular plastics.

Major parts of this Chapter are reproduced with permission of Science from:

Meys, R., Kätelhön, A., Bachmann, M., Winter, B., Zibunas, C., Suh, S. and Bardow, A. (2021). Achieving net-zero greenhouse gas emission plastics by a circular carbon economy. *Science*, 374(6563):71-76.

The author of this thesis developed the concept of the study, was in charge of the development of the set-up of the bottom-up model and methodology, wrote the first draft of the manuscript, and interpreted the results.

7.1 Achieving net-zero emission plastics

The results show that a recycling pathway, via mechanical and chemical recycling, reduces greenhouse gas emissions by 3.0 Gt CO_2-eq or 64% compared to the linear-carbon pathway (Figure 7.1). Up to 4.5 Gt CO_2-eq (96%) are reduced by a biomass pathway, where biomass uptake recycles CO_2, compensating the emissions primarily due to plastic waste incineration and production and fossil-based feedstock like naphtha. While plastic waste and biomass provide carbon and sufficient energy for conversion, carbon capture and utilization technologies require electricity with a low carbon footprint to reduce greenhouse gas emissions, mainly to produce hydrogen by water electrolysis (Artz et al., 2018). For the current global average carbon footprint of electricity (Ecoinvent, 2020), carbon capture and utilization would increase plastics' greenhouse gas emissions. However, by using electricity with the current footprint of wind power, commercialized carbon capture and utilization technologies for methanol and methane can reduce up to 4.4 Gt of CO_2-eq (94%). Overall, plastics solely based on either recycling, biomass utilization, or carbon capture and utilization fail to reach net-zero emissions even for wind-based electricity production. Assuming wind-based electricity supply, between 0.2 Gt and 1.7 Gt of CO_2 would have to be abated by negative-emission technologies, like direct air capture with CO_2 storage (Fuss et al., 2018), to turn plastics net-zero.

In the single technology pathways, three sources exist that prevent fully net-zero emission plastics due to residual greenhouse gas emissions: waste incineration, biomass supply, and renewable electricity supply. In the maximal recycling pathway, recycling rates are maximized, but all recycling processes produce residual wastes. These residual wastes are incinerated, leading to unavoidable greenhouse gas emissions even for maximal recycling rates. Additionally, waste incineration emits small amounts of non-CO_2 emissions, like carbon monoxide or methane, increasing residual greenhouse gas emissions for biomass utilization and carbon capture and utilization. Biomass cultivation emits non-carbon greenhouse gases like nitrous oxide that the CO_2 uptake cannot avoid. Finally, even in the most ambitious scenarios of the International Energy Agency (IEA), renewable electricity production is not entirely net-zero by 2050 with approx. 13.5 g CO_2-eq per kWh (International Renewable Energy Agency, 2019), leading to residual greenhouse gas emissions from electricity supply in the CCU pathway. Even wind-based electricity, i.e., electricity production with currently the lowest greenhouse gas emissions, is not fully net-zero (IPCC, 2014).

In contrast, a circular-carbon pathway optimally combines recycling, biomass utilization, and carbon capture and utilization reduce greenhouse gas emissions of plas-

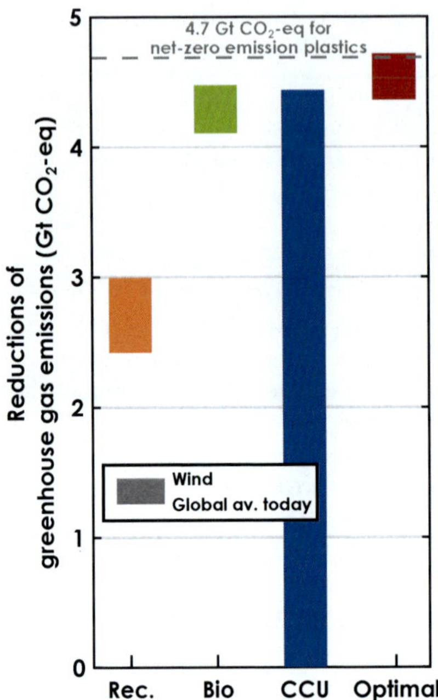

Figure 7.1: Reductions of life cycle greenhouse gas emissions from cradle-to-grave of the four pathways: recycling, biomass, carbon capture and utilization (CCU), and the circular-carbon pathway.

tics by up to 4.73 Gt CO_2-eq, assuming wind-based electricity (Figure 7.1). Thus, the plastics' life cycle would even be slightly net-negative. The shift is achieved since the combination of all circular technologies minimizes the residual greenhouse gas emissions (Figure C.1 in Appendix C.9 for a detailed Sankey diagram and carbon flows).

Biomass takes up CO_2 and thereby off-sets CO_2 emitted from waste incineration and production processes. The respective biomass is then gasified to generate synthesis gas, a mixture of hydrogen and carbon monoxide, for methanol production. CO_2 produced during biomass gasification is captured using established technologies, e.g., the Rectisol process (Appendix C). No direct air capture is used. The captured CO_2 from biomass gasification is directly converted to methanol using thermal hydrogenation. Subsequently, methanol is used to produce ethylene, propylene as well as benzene, toluene, and xylene, which are the primary raw materials for the 14 large-volume plastic materials. The related process technologies for methanol-to-olefins and -aromatics conversion are already industrialized. In 2018, 28% of the global methanol production was used for ethylene and propylene production (ICIS, 2018). Thus, producing renewable methanol from biomass and captured CO_2 offsets the primary source of residual greenhouse gas emissions in the recycling pathway: the incineration of residual wastes.

At the same time, recycling reduces the overall demand for biomass and renewable electricity and the corresponding residual greenhouse gas emissions. Two classes of recycling technologies achieve the reduction of the demand for biomass and renewable electricity: First, mechanical recycling is used for plastic packaging wastes to produce recycled resins. Each plastic resin produced by mechanical recycling does not have to be produced from biomass or CO_2. Second, plastics are treated by pyrolysis to produce naphtha feedstock. This naphtha feedstock is then used in steam crackers and solvent extraction processes to produce ethylene, propylene as well as benzene, toluene, and xylene. These chemical raw materials can be converted to plastics by conventional, industrialized technologies, reducing the need for biomass utilization or carbon capture and utilization technologies. As a result of recycling, the overall demand for biomass utilization or carbon capture and utilization technologies decreases, as do their residual greenhouse gas emissions.

Beyond the reduction of residual greenhouse gas emissions, part of the CO_2 taken up by biomass and carbon capture and utilization technologies is stored in landfilled plastics due to an unavoidable landfilling rate of 6%, based on Geyer et al. (2017). This permanent carbon storage compensates for the residual greenhouse gas emissions that still occur in the circular-carbon pathway. However, permanent carbon storage in landfills cannot be seen as sustainable since landfilling, managed or mismanaged, is the primary cause of plastic pollution (Gabrielli et al., 2020). Nevertheless, a certain amount of plastic leakage seems unavoidable, even in the most ambitious policy scenarios (Borrelle et al., 2020; Lau et al., 2020). A conservative assumption that the carbon in landfilled plastics entirely turns into CO_2 would lead to additional emissions of 0.24 Gt of CO_2-eq. In this case, the circular-carbon pathway would lead to net emissions of 0.21 Gt CO_2-eq.

The technologies and carbon feedstocks minimizing greenhouse gas emissions of plastics depend on the electricity supply's carbon intensity since different combinations of recycling, biomass, and carbon capture and utilization lead to varying synergies and electricity consumptions (Figure 7.2 upper part). As the carbon intensity of electricity decreases, the circular-carbon pathway with minimal greenhouse gas emissions employs more and more carbon capture and utilization technologies. This trade-off leads to break-even points for the carbon footprint of electricity supply at which carbon capture and utilization become climate beneficial compared to the recycling (39.5 g CO_2-eq per kWh) and biomass (5.8 g CO_2-eq per kWh) pathway. The electricity supply's carbon intensity also dictates the utilized carbon feedstock in the circular-carbon pathway (Figure 7.2 lower part): For electricity carbon footprints above 8.6 g CO_2-eq per kWh, the optimal pathway is solely based on biomass and plastic waste as carbon feedstock (Figure 7.2 lower part). Thus, these carbon inputs would be optimal for electricity impacts predicted for China, the United States, and Europe in 2040 that globally produce 57% of all plastics (ICIS, 2018; IEA, 2020). Thus, in all three major producing regions, the optimal combination of the circular-carbon pathways is based on biomass and recycling.

For electricity with carbon intensities below 8.6 g CO_2-eq per kWh, the carbon capture and utilization technologies become beneficial to minimize greenhouse gas emissions (Figure 7.2 lower part). If wind-based electricity is employed, the optimal circular-carbon pathway uses biomass, recycling, and carbon capture and utilization to produce 357 Mt of mechanically recycled and 814 Mt of virgin plastic. The virgin plastic production consumes 429 Mt of chemical feedstock from chemical recycling, 2148 Mt of biomass, and 949 Mt of CO_2. Conversion of the inert molecule CO_2 requires 9.9 PWh of renewable electricity (Figure 7.3).

Completely switching to CO_2-based products requires electricity with a carbon intensity below 6 g CO_2-eq per kWh. Today, not even wind power plants can provide such low-emission electricity (Schlömer et al., 2014). However, the switch from partly bio to entirely CO_2-based plastics reduces greenhouse gas emissions by only 1.5 %. Thus, using either biomass or CO_2 as carbon feedstock results in net-zero emission plastics, only if combined with large-scale plastic recycling.

7.2 Renewable resource demands

The analysis showed that combining carbon capture and utilization and/or biomass with large-scale recycling can achieve net-zero emission plastics. However, the actual feasibility will strongly depend on renewable resource availability. Here, two questions arise: Are there sufficient renewable resources to meet the global plastic demand? And, how does the circular-carbon economy perform against other pathways for net-zero emission plastics such as carbon capture and storage?

The circular-carbon pathway recycles 70% of the plastic waste back to plastics and

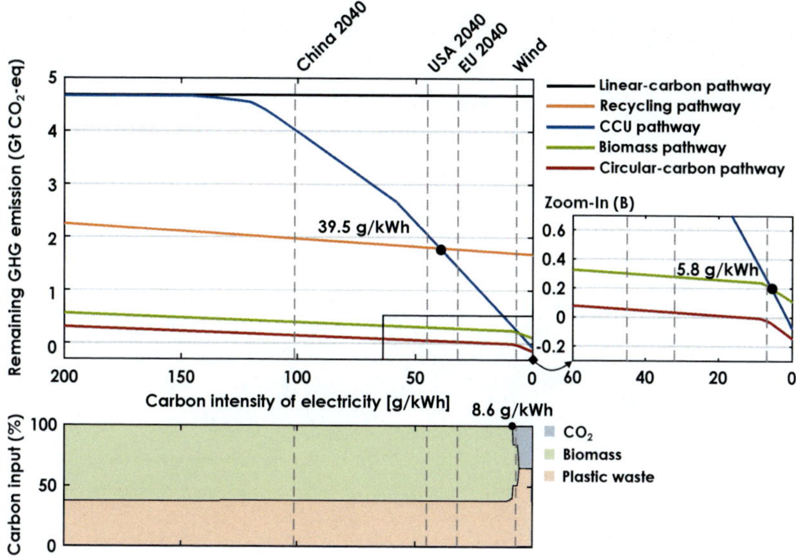

Figure 7.2: Remaining greenhouse gas emissions of the linear-carbon pathways and the four circular pathways depending on the carbon intensity of electricity and optimal carbon input in % of the circular carbon pathway depending on the carbon intensity of electricity. Note that the greenhouse gas emissions of the linear-carbon pathway do not alter due to a decrease of the electricity carbon intensity since all electricity is supplied from the energy recovery via plastic waste incineration and that the IEA report (IEA, 2020) only provides regionalized electricity impacts until 2040.

uses 19.3 EJ of biomass and 9.9 PWh of renewable electricity (Figure 7.4, black dot). Thus, the effective recycling rate of 70% is the maximum achievable due to losses in the recycling processes and the residual landfilling of 6%. The effective recycling rate includes only the actually recycled material. For instance, losses during mechanical and chemical recycling that ultimately are incinerated are not counted as recycled. The resource demands can be shifted between biomass and renewable electricity since both carbon capture and utilization and biomass can achieve net-zero emission plastics in combination with recycling. By increasing biomass supply and thus reducing the supply of CO_2 and vice versa, the electricity demand can vary between 1.6 and 18.1 PWh (Figure 7.4, red line). These electricity demands correspond to 59 to 670% of

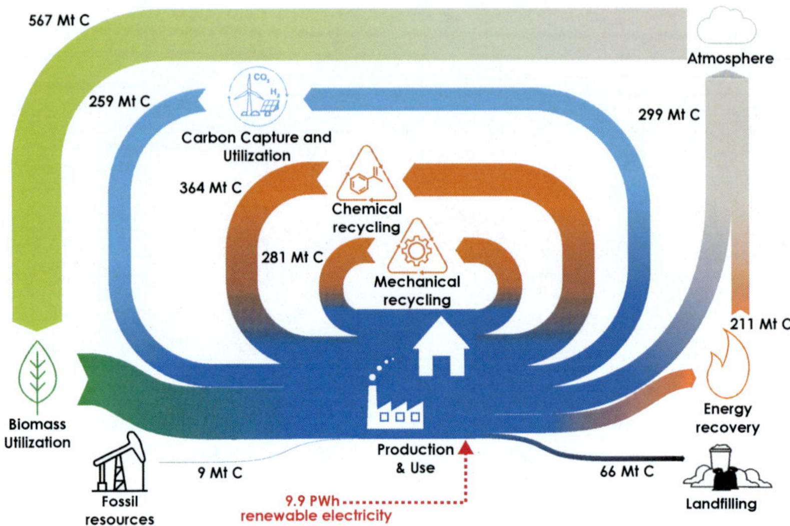

Figure 7.3: Feedstock supply and waste treatment of circular-carbon pathway for wind-based electricity production with 7 g CO_2-eq per kWh. The width and values represent the carbon content (Mt C) of the flows.

the electricity predicted to be supplied to the chemical industry (2.7 PWh) in the IEA's most ambitious scenario for 2050 (IEA, 2020). However, the electricity supply calculated by the IEA follows a cost-optimal demand in the petrochemical industry, to achieve a specific emission-reduction target. Thus, the calculated electricity supply rather approximates the electricity amount that will be supplied, and not that could potentially be supplied to the chemical industry. Furthermore, the plastics industry is not entirely covered by the IEA model, and thus, the electricity demands for plastics are not fully represented. However, approximately 80% of the overall mass of the major bulk chemicals (e.g., ethylene, propylene, benzene, toluene, and xylene) are used for plastics. Thus, the 2.7 PWh of electricity supplied to the chemical industry is assumed as a reasonable estimate for a possible amount of electricity that can be supplied to plastics life cycles. As a result, the combination of 2.7 PWh electricity with 42.6 EJ of biomass to achieve net-zero emission plastics is denoted as the "feasible point" (Figure 7.4).

Estimates of biomass availability vary widely from 30 EJ to over 1000 EJ per year

(Jones and Albanito, 2020). An expert consortium found that an estimate of 100 EJ is supported by "high agreement" in literature (Creutzig et al., 2015). The International Renewable Energy Agency (IRENA) estimates that an additional 287 EJ are available (International Renewable Energy Agency, 2019, 2016) by using lignocellulosic biomass and food wastes as well as land made available by technology and farming improvements. Considering the future biomass demand (104 EJ) of all other sectors (IEA, 2020), IRENA's estimated untapped biomass supply equals 183 EJ. Thus, the biomass demand of 42.6 EJ at the feasible point would represent 23% of this remaining untapped biomass potential. However, the access to this untapped biomass relies on boosting crop yield, reducing food waste, afforestation, and improving livestock management to free pastureland (International Renewable Energy Agency, 2019, 2016).

To put the renewable energy demands into perspective, they are compared to the linear-carbon pathway that would need 4.7 Gt CO_2 storage, e.g., using carbon capture and storage, to reach net-zero emission plastics. In this case, 76.9 EJ of fossil-based energy and between 1.9 and 33.9 EJ of additional electricity for carbon capture and storage would be required (Figure 7.4 grey area). The range of electricity reflects the impact of the CO_2 source from a high-concentration stream (e.g., ammonia (von der Assen et al., 2016)) to direct air capture (DAC) (Deutz and Bardow, 2021). The industrial-scale capture of CO_2 during ammonia and hydrogen production is well established and frequently performed by the Rectisol process considered here (Elvers and Ullmann, 2011; IHS Markit, 2018). For direct air capture, recently published data for the Climeworks plants in Hellisheioi and Hinwil is used, assuming wind-based electricity supply (Deutz and Bardow, 2021). Overall, the linear-carbon pathway plus carbon capture and storage would consume between 78.8 and 110.8 EJ of energy. Thus, the circular-carbon pathway at the feasible point with a total energy consumption of 42.6 EJ of biomass and 9.7 EJ of electricity, could potentially save between 34 and 53% of the total energy demand.

Energy demands can be further reduced by combining the recycling pathway with carbon capture and storage to achieve net-zero emission plastics (Figure 7.4, orange area). This combination requires 40.6 to 52.2 EJ while achieving net-zero emission plastics, reducing energy by an additional 1 to 12% compared to the circular-carbon pathway. Thus, the total life cycle energy demand differs by up to 12% between carbon capture and storage, biomass, and carbon capture and utilization. However, in all cases, recycling plastic waste reduces the energy demand compared to the fossil-based benchmark with carbon capture and storage.

While the decreased energy demand may appear counter-intuitive due to the anticipated lower efficiency of biomass and CO_2-conversion, it can be rationalized by energy conservation over the complete life cycle and recycling: The pathways based on fossils, biomass, and CO_2 can recover the energy contained in plastics only during waste incineration. Energy recovery is inefficient due to unavoidable losses from thermal energy to electricity conversion due to thermodynamic limitations. Therefore, plastic waste incineration will inevitably never suffice to close the energy loop and maintain 100% of the energy content. In contrast, recycling essentially conserves the energy content of plastics by re-utilizing plastic waste and avoiding its incineration, and thus lowering energy demands. A major difference is that the linear pathways can exploit fossil energy generated over millennia, while the biomass and CCU pathways need to generate their energy now.

To achieve these reduced energy demands, the circular-carbon economy can rely on commercialized recycling technologies. All technology datasets used in the main paper are based on already commercialized technologies using industrially validated data or detailed process simulations. Sorting household plastics and mechanical recycling is already industrial practice in western European countries like Germany or Austria (Reichel et al., 2016). While plastic packaging can be efficiently recycled mechanically, mixed and other plastic wastes lack mechanical recyclability (Kümmerer et al., 2020; Ragaert et al., 2017). Here, pyrolysis offers a promising large-scale avenue to increase recycling rates (Jeswani et al., 2021).

Additionally, early-stage technologies converting plastic waste to their respective monomers are currently under development (Hong and Chen, 2017). To highlight the potential benefits of technology development, emerging technologies are included in Appendix C.10. By leveraging potential low-TRL chemical recycling, biomass utilization, and carbon capture and utilization technologies, the circular-carbon pathway could reduce the energy demand even by 83% compared to the linear-carbon pathway with carbon capture and storage (Figures C.2 and C.3). The most promising low-TRL chemical recycling technologies are those that substitute fossil-based chemicals associated with high greenhouse gas emissions (cf. also Chapter 3) and for which currently no mechanical recycling exists. In particular, chemical recycling technologies producing isocyanates and polyols from polyurethanes, styrene from polystyrenes, caprolactam from polyamide 6, or adipic acid and hexamethylenediamine from polyamide 66. Furthermore, it can be observed that the shift from biomass to CO_2 utilization starts at a carbon intensity of electricity supply of 134 g CO_2-eq per kWh. In particular, the early-stage technologies to produce ethylene and propylene, as well as benzene, toluene and xylene are utilized. Note that a further detailed analysis of early-stage carbon capture and utilization technologies is provided in Appendix D.

However, even promising recycling technologies with low technology readiness level

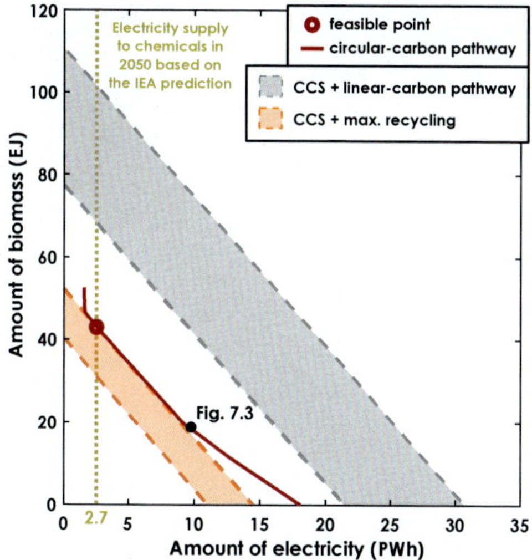

Figure 7.4: The circular-carbon pathway is shown together with the linear-carbon and recycling pathways with carbon capture and storage. The range of carbon capture and storage reflects different sources of CO_2, such as ethylene oxide plants (von der Assen et al., 2016) or ambient air (Deutz and Bardow, 2021). The energy demands for the linear-carbon pathway are based on fossil resources, which are converted to biomass and electricity on an energy basis.

can only leverage their potential if sufficient plastic waste is collected and made available. A recent study predicted for 2040 that 88% of the plastic demand will still be lost in managed landfills and waste incinerators (both 32%) or through waste mismanagement (56%) (Lau et al., 2020). To increase collection rates, landfill bans have been proven highly effective in European countries like Germany. In Europe, landfilling dropped by 34%, while recycling increased by 64% between 2006 and 2014 (PlasticsEurope, 2018). Once the collected waste is in a managed system, recycling can be fostered by implementing recycling quotas. However, a proper waste management infrastructure is missing in many low- and middle-income countries that would offer the largest potential to access plastic waste (Borrelle et al., 2020; Jambeck et al., 2015). To gain access to these untapped resources, millions of households must be connected to waste management services. Lau et al. (2020) argue that this monumental task requires linking local service chains (e.g., the informal sector (Kaza et al., 2018)) to the value chain (e.g., recycling) by increasing the profitability of material recycling through investments in waste management infrastructure and improved coordination for collection, sorting and management of plastic wastes (Lau et al., 2020).

7.3 Operational cost

The operational costs differ between the linear and circular-carbon pathways in two major aspects: (1) the costs for feedstock and energy to produce plastics and (2) the costs for the end-of-life treatment of post-consumer plastics, e.g., landfilling, energy recovery, and chemical and mechanical recycling.

Taking into account the expected price ranges (Section C.11) for biomass, CO_2, electricity, and oil (IEA, 2018b, 2020) as well as the operational cost for mechanical (Rudolph et al., 2017) and chemical recycling (IHS Markit, 2018), and waste incineration (Rudolph et al., 2017), the operational costs of the circular-carbon pathway vary between 822 and 1366 billion USD on a global basis in 2050 (Table 7.1). Thus, compared to the linear-carbon pathway with operational costs between 839 and 1110 billion USD, the circular-carbon pathway lies in the same cost range. Assuming costs at the low end for all resources would lead practically identical cost for the linear-carbon pathway (839 billion USD) and the circular pathway (822 billion USD). For high oil prices around 70 USD per barrel as well as a low-cost supply of biomass, CO_2 and renewable electricity with approx. 5 USD per GJ, 30 USD per ton CO_2 and 2 USD per kWh, the circular carbon pathway would even save 288 billion USD in operational cost. Depending on the resource prices, the respective CO_2-abatement of 4.7 Gt CO_2-eq would cost -61 to 112 USD per kg of CO_2-eq abated. This range shows considerable uncertainty due to the broad price ranges for oil, biomass, CO_2, and electricity. However, even the high-end carbon cost of 112 USD per kg of CO_2-eq is in the predicted range of prices for CO_2 certificates in Europe or mid-century social costs of carbon (Fragkos, 2011; Ricke et al., 2018), indicating that, with regard to operational costs, current carbon price scenarios would suffice to reach net-zero emission plastics.

The abatement costs are based only on the operational costs and do not include capital expenditures. However, an increase in capital expenditures can be expected to provide plastic, biomass, or CO_2 as a resource. These investments are needed for the infrastructure to provide the renewable carbon feedstock to the production facilities of the chemical industry and to convert the renewable feedstock to plastic products. While a full assessment of capital expenditures is not possible due to the lack of data in this thesis, the increase can be estimated based on a recent calculation by the IEA. To reduce the greenhouse gas emissions of the petrochemical sector, which is closely connected to the plastic industry (Levi and Cullen, 2018), the IEA estimated an investment need of 1.5 Trillion USD (IEA, 2018a) to save 0.9 Gt CO_2-eq yearly after 2050. While these investment amounts do not fully represent those required to achieve net-zero emission plastics, they clearly show that the additional capital expenditures will increase abatement costs. Assuming the same capital expenditures as the IEA, an average production plant lifetime of 30 years, and 0.9 Gt CO_2-eq savings for these years, the abatement cost would increase by approx. 56 USD per ton of CO_2.

Table 7.1: The operational costs are shown for oil, biomass, CO_2, and electricity and the amount of waste treated by mechanical recycling, chemical recycling, and energy recovery from left to right. Prices are given in Appendix C.11.

Resource / Waste volume	Oil	Biomass	CO_2	Electricity	Mech. recycling	Chem. recycling	Energy recovery	Total
Unit	billion USD	billion USD	billion USD	billion USD	billion USD	billion USD	billion USD	billion USD
low prices linear-carbon	675						164	839
high prices linear-carbon	946						164	1110
low prices circular-carbon	18	212	1	54	82	413	42	822
high prices circular-carbon	25	639	3	162	82	413	42	1366

7.4 Policies for circular and net-zero emission plastics

The results indicate that net-zero emission plastics can be achieved using technologies already available and commercialized today. To materialize the potential, the design and implementation of policies are needed that foster the deployment of circular-carbon technologies. This thesis identified two crucial technological changes to achieve net-zero emission plastics: (1) increase plastic recycling rates and supply more plastic waste feedstock and (2) deploy carbon capture and utilization or biomass technologies, depending on local availabilities of renewable electricity and biomass. Fostering such changes requires economic incentives.

Economic incentives can play a crucial role in increasing plastics' circularity and achieving net-zero emission plastics. However, under the current structure, pricing carbon emissions would have at best-limited impact in incentivizing plastic circularity. Current emissions trading schemes (ETSs) focus on production processes and exclude end-of-life (EOL) management processes (European Commission, 2010) such as incineration from the scope, leading to a problem shifting from one stage of the life cycle to another. Including plastic waste incineration in emission-pricing schemes through, e.g., extended producer responsibility (EPR) policies, would be a step forward to incentivize recycling (Leal Filho et al., 2019). For carbon pricing to effectively reduce greenhouse gas emissions throughout the plastics' life cycle and improve the circularity of plastics, it is crucial to cover the entire life cycle of plastics within the scope of carbon pricing.

Additionally, local municipalities often conduct waste management as a service, particularly collection and sorting, for the residents or local companies. Since the residents pay for the service, only limited incentives exist to generate a usable waste stream. In return, municipalities often receive post-consumer waste of low value that is poorly usable. These post-consumer wastes are then treated at the lowest potential costs, e.g., waste incineration or, even worse, landfilling, instead of proper sorting. Thus, collection and sorting are commonly the bottlenecks of recycling industries (Ragaert et al., 2017). Policymakers should aim to incentivize value addition at the beginning of the waste value chain: the plastic consumer. Here, a deposit system for plastic materials could provide a potential avenue for consumers to provide valuable plastic waste feedstock directly from the start. These deposit systems have been very successful in European countries like Germany for the use case of plastic drinking bottles (Picuno et al., 2021).

While increasing the availability of plastic waste is one important point to be addressed by policies, current policies also subsidize oil exploration and production of fossil products and thereby offer a cheap and abundant alternative to plastic waste, biomass, and CO_2 as renewable carbon feedstocks. In the oil and gas industry, investments in the extensive infrastructure have already paid off. In contrast, plastic waste, biomass, and CO_2 utilization are only in their infancy. Thus, investment possibilities currently do not satisfy private investors due to lower return-on-investments. As a result, the initial capital investment largely disincentives potential investors, particularly in the recycling area (Gao et al., 2021). However, investment firms like Black Rock are increasingly interested in including climate impact in their investment decisions (Black Rock, 2021).

Thus, the use of globally agreed policies (Simon et al., 2021) could increase the availability of plastic waste and incentivise investment in biomass and CO_2 utilization. Looking ahead, the circular carbon economy could hold its promise to redesign plastics production systems such that decoupling from fossil carbon resources achieves net-zero emission plastics with lower energy demands while requiring reduced operational costs and, thus, combining economic and environmental well-being (Foundation, 2016).

Chapter 8

Summary, discussion, and perspectives

Plastics are loved and hated but are undoubtedly on the rise to conquer every area of modern human life (OECD, 2018; Foundation, 2016). However, this rise comes at the costs of large-scale environmental burdens, like ever-increasing plastic pollution of natural environments (Haward, 2018; Isensee and Valdes, 2015), an enormous share of global oil consumption of 20% by 2050 (IEA, 2018a), and, finally, 15% of the yearly-allowed greenhouse gas emissions in 2050 to keep global warming below 1.5°C (Foundation, 2016). Thus, combating climate change and meeting global climate targets requires net-zero greenhouse gas emission plastics by the second half of this century (Rogelj et al., 2018; Zheng and Suh, 2019).

To reduce the greenhouse gas emissions of plastics, three circular technologies can be used (Bazzanella and Ausfelder, 2017; Zimmerman et al., 2020): (1) chemical or mechanical recycling, (2) carbon capture and utilization, and (3) biomass utilization. Current scientific literature focuses on assessments of individual or partly combined circular technologies and limited regional scope (Hermann et al., 2007; Artz et al., 2018; Zheng and Suh, 2019; Schwarz et al., 2021; Posen et al., 2017; Saygin et al., 2014), often applies inconsistent methodologies (Laurent et al., 2014a; Lazarevic et al., 2010; Spierling et al., 2018), or focuses on reductions and comparative assertions rather than how net-zero greenhouse gas emissions can actually be achieved (Finkbeiner and Bach, 2021). Due to the focus of literature, a comparable, systematic, and industry-wide assessment of circular technologies to enable net-zero emission plastics is, thus, missing. Additionally, the transition from a linear to a circular economy is frequently rated as too energy-intensive and costly, leaving policymakers with an unclear basis to make sound decisions for a net-zero transition.

To close the gap in literature, this thesis builds and uses the first global and industry-wide bottom-up model for plastic production and waste treatment. The bottom-up model represents the global greenhouse gas emissions of 90% of the global plastic production volume. Based on this bottom-up model, the thesis highlights that combining all three circular technologies can achieve net-zero emission plastics while consuming less energy and, under favorable market conditions, requiring less operational costs than the fossil-based benchmark with large-scale carbon capture and storage.

The following Section 8.1 summarizes the conclusions of this thesis. In the subsequent Section 8.2, the model validity is discussed, and an outlook for future research is given.

8.1 Summary and conclusions of this thesis

During the literature review of this thesis (Chapter 2), the following gaps and challenges have been identified: (1) many circular technologies in early development stages have to be evaluated against their conventional benchmarks despite low data availability, (2) even though several mature technologies exists, the literature lacks technical data and life cycle assessments of the plastics supply chain, and (3) if sufficient data would be available, this data needs to be combined in a consistent and industry-wide model to systematically evaluate if and how net-zero emission plastics can actually be achieved.

To close the respective gaps in literature, this thesis first assesses early-development chemical recycling for plastic packaging wastes from an environmental perspective, despite low data availability (Chapter 3). To do so, ideal thermodynamic relations are used to calculate mass- and energy balances of chemical recycling technologies, leading to minimized environmental impacts. These minimal environmental impacts are then compared to the respective benchmark waste treatment technologies based on industry data. If the minimal environmental impacts of the ideal chemical recycling exceed those of the benchmark waste treatment, the results are robust since more realistic conditions for chemical recycling will only lead to higher environmental impacts. Overall, this thesis shows that it will be very challenging for chemical recycling to compete with mechanical recycling. In contrast, substituting waste incineration offers the potential to reduce greenhouse gas emissions for all chemical recycling technologies.

Even though the methodology focuses on recycling technologies, the general approach of using ideal thermodynamics can be adapted to any novel early-stage circular technology. The environmental impacts of novel technologies, particularly greenhouse gas emissions, are driven by the respective raw material and energy use. The raw material and energy uses are minimal if ideal thermodynamic conditions are assumed.

Despite its flexibility, the method is limited such that only the most promising circular technologies can be identified. However, reductions of environmental impacts do not imply an actual benefit by the circular technology compared to its benchmark. To ensure environmental benefits, a full life cycle assessment is warranted. Thus, the methodology developed in this thesis is limited to robustly identify those circular technologies, and in particular chemical recycling technologies, that are inferior to the benchmark even for ideal conditions.

A full life cycle assessment is performed in Chapter 4 for the case of CO_2-based rubbers. Here, the environmental impacts of CO_2-based rubbers are compared with the conventional benchmark products, e.g., nitrile- and ethylene/propylene-based rubbers. The analysis, which is based on data from pilot plants for CO_2-based rubbers, shows that that CO_2-based rubbers can reduce between 1.1 and 2.5 kg of CO_2-eq emissions per functional unit compared to their benchmarks. This reduction is mainly due to the substitution of conventional rubbers and reduced emissions from the end-of-life incineration. However, the life cycle assessment also revealed that other environmental impacts like stratospheric ozone depletion or marine eutrophication are increased by 1 to 54%. This increase is due to the utilization of the energy-intensive reactant propylene oxide.
Additionally, the increases highlight the potential trade-offs between reductions of greenhouse gas emissions and increases in other environmental impacts. While applying life cycle assessment to CO_2-based rubbers adds one case study to scientific literature, the study also underlines that 1 kg of CO_2-based rubber still leads to net-positive greenhouse gas emissions between 4.4 and 7.0 kg of CO_2-eq depending on the CO_2-based rubber composition. 56.7% of these greenhouse gas emissions are due to the supply of still partly fossil-based chemical raw materials, particularly propylene oxide. As a result, the case study highlights the need for a more global and holistic assessment of complete supply chains from chemical feedstock over chemicals to plastic materials to evaluate potential pathways towards net-zero emission plastics.

The need for such a global and holistic assessment is addressed by building the global bottom-up model of plastics life cycles, described in Chapter 5. The bottom-up model itself as well as the subsequent applications to several case-studies (Chapters 6 and 7 as well as Appendices D and E) address all scientific gaps addressed in this thesis.

The bottom-up model is based on over 400 technical datasets representing the production, consumption, and end-of-life (e.g., from cradle-to-grave) of 75 chemicals and 14 plastic materials. As a result, the model covers 90% of the global plastic production volume, at least 75% of greenhouse gas emissions, and 60% of energy consumption by the (petro)chemical industry (Ausfelder et al., 2013). The datasets used to represent the supply chain of plastics have been generated based on scientific and peer-reviewed literature, a standard LCA database (Ecoinvent, 2020) and detailed process simulations (IHS Markit, 2018). See Appendix C for a detailed list of all literature sources. Thus, the data quality of the bottom-up model can be seen as high. After compiling and harmonizing the datasets, it is included in the computational structure of the Technology Choice Model (Kätelhön et al., 2016) that enables the optimization of the greenhouse gas emissions of the complete plastic supply chain.

The validity of the bottom-up model is evaluated against the current status of literature for the year 2015 (cf. Chapter 6). It is found that the utilization ratios of chemicals for plastic materials only vary by approx. 4% on average compared to ICIS (2018), the overall energy consumption calculated differs by only 5% from calculations by the International Energy Agency (IEA, 2018b), and the greenhouse gas emissions differ by 8% compared to recent estimates by Zheng and Suh (2019). Thus, these small differences indicate that focusing on the 14 plastics, the best available fossil technologies, and using an optimization algorithm to derive the full plastics supply chain represents a good compromise between model accuracy and additional modeling effort. A discussion about potential improvements and further research is provided in Section 8.2.

In a final contribution to the scientific literature, this thesis uses the bottom-up model to evaluate feasible pathways to achieve net-zero emission plastics via the three circular technologies (cf. Chapter 7) and addresses the gap 3 identified in this thesis. In particular, greenhouse gas emissions of plastics from cradle-to-grave are calculated for the year 2050 and five pathways. These pathways include a linear and fossil-based pathway, a maximal recycling pathway, a carbon capture and utilization pathway, a biomass pathway, and a circular-carbon pathways that combines all three circular technologies.

The results show that applying a single circular technology fails to achieve net-zero emissions plastics due to the occurring residual greenhouse gas emissions. These residual greenhouse gas emissions are based on (1) residual waste during recycling that ultimately has to be incinerated, (2) nitrous oxides and methane emissions during biomass cultivation that cannot be avoided via the CO_2 uptake, and (3) residual greenhouse gas emissions due to renewable electricity supply. In contrast, the analysis shows that net-zero emission plastics can be achieved by optimally combining all three circular technologies: The recycling of plastic wastes reduces the overall demand for biomass and renewable electricity and, thus, the respective residual greenhouse gas emissions. Additionally, biomass and CO_2 utilization recycle CO_2 from the waste incineration of residual wastes from all recycling processes and, thereby, abate the respective residual emissions.

Furthermore, the analysis highlights that the synergies of all circular technologies lead to up to 53% lower energy demand of the circular-carbon pathway compared to the linear-carbon pathway with large-scale carbon capture and storage. While these lower energy demands seem counter-intuitive, they can be rationalized by the energy conservation of chemical feedstock via chemical and mechanical recycling. The analysis of circular technologies finishes with an estimation of operational costs. Assuming low costs for biomass, renewable electricity, and CO_2, and high costs for fossil resources, e.g., crude oil, this thesis shows that due to the reduced energy demands, the operational costs can be 288 Billion USD lower than the linear-carbon pathways. However, for the unfavorable market conditions, the circular technologies would increase 527 Billion USD of operational costs. Thus, overall, this thesis shows that net-zero emission plastics are achievable with lower energy demands based on the existing commercialized circular technologies. However, economic benefits strongly depend on favorable market conditions.

These favorable market conditions need policy actions to foster the transition towards circular and net-zero emission plastics. These policies include, not exclusively, (1) consistent greenhouse gas emissions pricing without neglecting parts of the life cycle, e.g., waste incineration, (2) introducing deposit systems for many plastic products in order to incentivize the generation of a valuable waste stream directly from the start, e.g., the plastic consumer, and (3) stop subsidizing fossil resource production in order to increase competitiveness of renewable resources and their utilization.

Using these policy instruments to increase the availability of plastic waste and to provide economic incentives for increased investment in biomass and CO_2 utilization can foster the transition towards net-zero emission plastics. However, the improvements regarding energy consumption and greenhouse gas emissions have to be carefully balanced with other environmental impacts known to arise due to the large-scale usage of biomass or renewable electricity for CO_2 utilization, for which some examples are shown in Chapters 3 and 4 of in this thesis. These environmental impacts, for instance, include an increase in terrestrial acidification and water eutrophication due to biomass utilization (Walker and Rothman, 2020) or the increase of metal depletion and water consumption due to the increase of solar and wind-based power plants (Garcia-Garcia et al., 2021). Then, the circular-carbon economy could hold its promise to redesign the man-made production systems such that the decoupling from fossil carbon resources achieves net-zero emission plastics with lower energy demands and climate impacts while requiring less operational costs and, thus, combine economic and environmental well-being (Foundation, 2016).

Thus, this thesis shows that the greenhouse gas emission problem of plastics can be solved. All technologies and solutions are available, and all that needs to be done is wake up and change.

8.2 Discussion and perspectives for future research

While the last section summarizes the contributions and conclusions of this thesis, this section focuses on a discussion of the model validity and perspectives for future research.

The comparison of the bottom-up model to the current calculations in the literature, revealed smaller differences that can be attributed to particular assumptions during the generation of the bottom-up model.

First, the model only covers 90% of plastic production and excludes several smaller scale plastic materials like elastomers or high-performance products (Wypych, 2012). Thus, utilization shares of the high-volume chemicals (cf. Figure 2.4) for other plastic materials differ between the model and the reality. As a result, these applications are not represented by the model.

Second, the bottom-up model focuses on best-available, fossil-based production technologies. Thus, the production of ethylene and propylene from coal-based methanol that is mainly practiced in China has not been included. As a result, the energy demands of the bottom-up model currently do not depict the actual utilization shares in every detail.

Third, the optimization approach minimizes the greenhouse gas emissions of plastics. This minimization leads to lower greenhouse gas emissions than recent estimates by Zheng and Suh (2019) and will also rather underestimate actual greenhouse gas emissions for future scenarios, including the circular-technology pathways. This underestimation is due to market effects that would, at least in some cases, lead to suboptimal decisions by stakeholders (Kätelhön et al., 2016). Furthermore, in reality, decisions about which technology is used to produce which chemical and plastic are based on economics and not a minimization of greenhouse gas emissions.

To overcome these challenges several improvements could be made during future research. The first improvement could be an inclusion of smaller scale plastic applications, implementing additional fossil-based technologies, particularly the coal-based chemistry in China, or even completely other sectors like cement and steel. By this means, the already existing applicability of the bottom-up model to chemical and plastic production can be further broadened to specialty chemicals and plastics or the assessment of the industrial symbiosis within the energy-intensive industries, including chemicals, plastics, cement, and as steel and iron.

In addition to the extension above to cover more technical detail, it is also important to evaluate other environmental impacts. The bottom-up model is already able to cover a multitude of environmental impacts, like eutrophication, metal depletion, or acidification and, thus, should be used to assess these other environmental impacts. Another potential avenue is to connect the model with the planetary boundary concept (Galán-Martín et al., 2021) and evaluate whether net-zero emission plastic will stress the other environmental impacts.

Furthermore, the analysis in Chapter 7 revealed the lack of capital expenditures in assessing the economic feasibility of the circular-carbon pathway. However, both investment costs and further operational costs, like labor costs, are already available for many detailed process simulations used as background data (IHS Markit, 2018). By including both capital expenditures and full operational costs, the objective function could be extended or exchanged to represent the economics of the respective circular technologies and the respective transition to a more sustainable future. Based on an economic objective, several methodological extensions would be possible, including dynamic modeling of transition pathways or market-mediated effects (Kätelhön et al., 2016), e.g., rebound effects. Based on the inclusion of dynamic modeling and market-mediated effects, the bottom-up model could be further extended to scenario modeling of feasible transition pathways from today to 2050 under certain policy scenarios, for instance, the inclusion of full emission pricing schemes (European Commission, 2010).

While all improvements above focus on extending the bottom-up model to other applications and objectives, it is important to tailor the model to more explicit uses. One of such uses is an increased regionalization of the inherent bottom-up approach. For instance, the general bottom-up model structure could build a digital twin of integrated production sites, like Ludwigshafen of the BASF, and derive pathways for these productions sites to achieve net-zero emission targets. By this means, the bottom-up model would be much more applicable to actual use-cases with more practical relevance for stakeholders. With this regard, the bottom-up model could be extended to model company-wide reduction targets or derive regional roadmaps for chemical production sites and entire regions or continents.

Furthermore, the mathematical structure of the model could be used as a full life cycle assessment modeling tool based on primary data from actual production facilities. For this purpose, the technology data would need to be filled from enterprise resource planning (ERP) software applications. These ERP systems include mass and energy balances or bills of materials that can be incorporated into the bottom-up model structure to enable calculate environmental impacts of a large number of potential chemical and plastic products.

Finally, on a more practical level, the approach to process large amounts of technical data can be combined with market intelligence and trade data to build a fully regionalized and fully life-cycle-assessment-compliant bottom-up model of the chemical and plastic industry. In fact, the work conducted in this thesis is the foundation of the company Carbon Minds (CarbonMinds, 2020) which will leverage the developments during this thesis.[1]

Thus, the work conducted during this thesis will be beneficial for science and industry by increasing the transparency about the environmental impacts of chemical and plastic supply chains while enabling policymakers to shape the transition towards net-zero emission plastics.

[1]Raoul Meys is co-founder, CTO and managing director of Carbon Minds.

Appendix A

Supplementary information Chapter 3

A.1 Composition of plastic packaging waste

Table A.1: Composition of plastic packaging waste after sorting plants (HTP GmbH & Co. KG, 2017; DSD, 2018)

content type	unit	HDPE	LDPE	PP	PET	PS
polymer	kg	0.846	0.828	0.846	0.882	0.846
moisture	kg	0.100	0.100	0.100	0.100	0.100
residuals (metals, paper etc.)	kg	0.054	0.072	0.054	0.018	0.054

A.2 Modeling of energy recovery

The system boundary for energy recovery is shown in Figure A.1. Direct emissions of energy recovery are modeled according to Doka (2003, 2013). The Doka model represents the current model for waste incinerators with energy recovery in ecoinvent (Ecoinvent, 2020). The model is based on the elementary composition of each waste fraction. Based on transfer coefficients for each element, e.g. C, H, N, O, the elementary flows (e.g. emissions) to air, soil or water are calculated. For each element, the model includes material requirements, such as washing agents or natural gas for heat, based on measured data from Swiss waste incinerator plants. The heating value for each waste input is used to calculate the energy output. For European municipal solid waste incinerators, the total energy efficiency is reported with 41% and a power-to-heat ratio of 0.35 (Eriksson and Finnveden, 2009). In case of energy recovery in cement kilns, lignite, natural gas and biomass are substituted based on equal net calorific values and equal thermal energy efficiency. Internal heat and electricity consumption for flue-gas cleaning is assumed equal to municipal waste incinerators (Doka, 2003, 2013). All datasets for avoided environmental impacts of energy production are obtained from ecoinvent with the exception of biomass for heat generation, which is taken from Gerssen-Gondelach et al. (2014). District heating is modeled including heat supply based on 44% natural gas and 56% other sources, e.g. biomass, oil and lignite (Connolly et al., 2014).

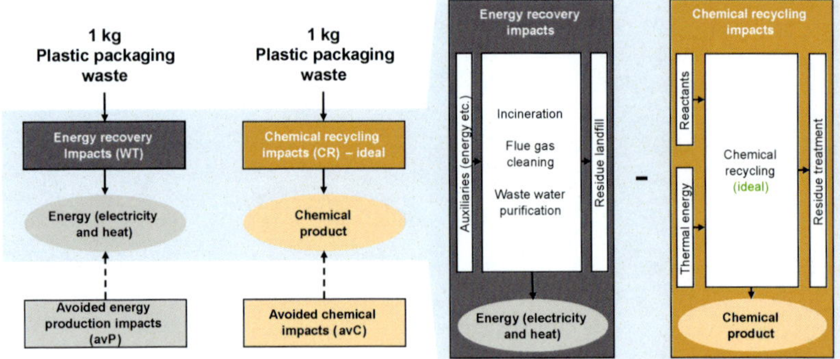

Figure A.1: Comparison between chemical recycling and energy recovery. The right side of the figure gives insight into the system boundaries and modeling for energy recovery and chemical recycling.

A.3 Dataset of energy recovery

Table A.2: Datasets for energy recovery in municipal solid waste incinerators of plastic packaging waste. Efficiencies are assumed 41% with a power to heat ratio of 0.35 (Eriksson and Finnveden, 2009).

Type	process or type	unit	HDPE	LDPE	PP	PET	PS
outputs	heat	MJ/kg	10.653	10.416	10.169	5.560	9.994
	electricity	MJ/kg	3.873	3.781	3.685	1.893	3.617
wastes	waste water to treatment	kg/kg	0.100	0.100	0.100	0.100	0.100
	waste residues to energy recovery	kg/kg	0.054	0.072	0.054	0.018	0.054

Table A.3: Datasets for energy recovery in cement kilns of plastic packaging waste. The heating value of lignite is assumed 23.64 MJ per kg (Ecoinvent, 2020).

Type	process or type	unit	HDPE	LDPE	PP	PET	PS
substituted lignite		MJ/kg	37.73	36.93	36.09	20.48	35.50
wastes	waste water to treatment	kg/kg	0.100	0.100	0.100	0.100	0.100
	waste residues to energy recovery	kg/kg	0.054	0.072	0.054	0.018	0.054

Table A.4: Specific global warming impact per kg of pure polymer calculated with the Doka model.

		unit	HDPE	LDPE	PP	PET	PS
composition	C	wt. %	85.63	85.63	85.63	62.50	92.30
	H		14.37	14.37	14.37	4.20	7.70
	O					33.30	
global warming impact		kg CO_2-eq	3.126	3.126	3.126	2.281	3.369

A.4 Modeling of mechanical recycling

The modeling of mechanical recycling is based on industry data from the year 2017 (DSD, 2018). Datasets represent the current technology standard for plastic packaging waste recycling.

Mechanical recycling requires sorted and to some extent pure plastic packaging waste to enable the production of high-purity polymer granulates (Rahimi and García, 2017). The sorting facilities are capable of sorting plastic packaging waste according to a given specification, such as the polymer or contamination contents. For example, a HDPE fraction with 94 wt. % HDPE, based on a dry mass basis, can be generated (DSD, 2018). The sorted fractions include 10 wt. % of moisture. The composition of all sorted fractions are given in Table A.1. The system boundaries for mechanical recycling are shown in Figure A.2 and are compared with those of chemical recycling.

Mechanical recycling includes a wet-processing pretreatment and extrusion step (HTP GmbH & Co. KG, 2017). The wet-processing consists of crushing the plastic packaging waste resource into smaller pieces, a scrubbing section to wash off organic and other contaminants and a density sorting to extract impurities of other plastic packaging wastes. The processes mainly require electricity supply from the electricity grid mix. To a smaller extent, flocculants are required to filter organic contaminants. Residual wastes comprise waste water, organic, metallic and paper contaminants as well as a fraction of the plastic that is lost during wet-processing and cannot be recovered for extrusion (HTP GmbH & Co. KG, 2017). The datasets for electricity supply, auxiliary materials (e.g. flocculants) and residual waste treatment are taken from the ecoinvent database (Ecoinvent, 2020).

During extrusion, the plastics are re-heated, melted and granulated and, thus, are exposed to mechanical and thermal stress resulting in their degradation. Thus, one kg of recycled granulate cannot fully substitute one kg virgin polymer granulate. A common method in life cycle assessment to take into account degradation is a so-called substitution factor. The substitution factor represents the ratio to which extent recycled polymer granulate can substitute virgin material. Substitution factors for all re-granulated materials are based on data provided by plastic recycling professionals are assumed (cf. Table A.5) (BIO Intelligence Service, 2013; Prognos AG, 2008). The avoided product emissions are calculated based on the environmental impacts of the virgin granulate production multiplied by the substitution factors.

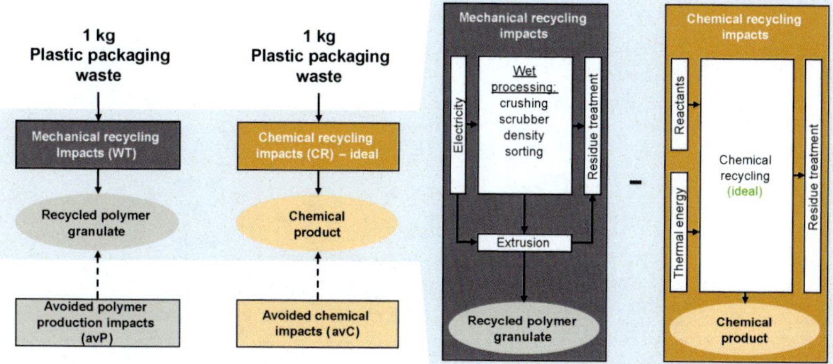

Figure A.2: Comparison between chemical recycling and mechanical recycling. The right side of the figure gives insight into the system boundaries and modelling for mechanical recycling and chemical recycling.

A.5 Datasets for mechanical recycling

Table A.5: Datasets for mechanical recycling of plastic packaging waste (HTP GmbH & Co. KG, 2017).

Type	process or type	unit	HDPE	LDPE	PP	PET	PS
treatment steps	wet processing (incl. crushing, scrubbing, density sorting)	kWh/kg	0.185	0.331	0.185	i	0.185
	treatment of waste water	kWh/kg	0.024	0.022	0.024	i	0.024
		process water in kg/kg	2.000	2.000	2.000	2.000	2.000
		NaOH in kg/kg	-	-	-	0.015	-
	extrusion (incl. drying)	kWh/kg	0.326	0.380	0.326	0.500	0.326
	air treatment and bio filter	kWh/kg	0.012	0.023	0.012	i	0.012
wastes	waste water to treatment	kg/kg	2.100	2.100	2.100	2.115	2.100
	waste polymer to energy recovery	kg/kg	0.093	0.075	0.093	0.130	0.093
	waste residues to energy recovery	kg/kg	0.054	0.072	0.054	0.033	0.054
product	secondary material	kg	0.75	0.75	0.75	0.75	0.75
	additional HDPE material	kg	-	-	-	0.050	-
substitution factor (sf) [1],[2]	recycled material = sf*virgin material		0.7	0.7	0.7	1	0.9

i Included in extrusion.

[1] BIO Intelligence Service (2013), [2] Prognos AG (2008)

A.6 Life cycle assessment results for benchmark waste treatment

In Figure A.3, the global warming impact is presented, while Figure A.4 contains results for the fossil resource depletion.

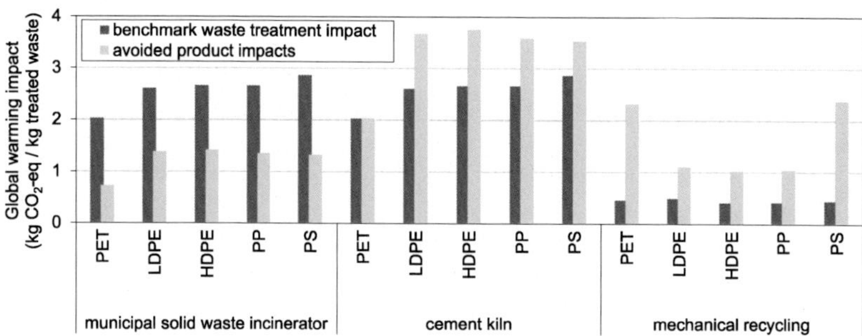

Figure A.3: The direct global warming impacts of benchmark waste treatments (dark grey) and credit for avoided products (light grey) are shown. All plastic packaging wastes are shown for each benchmark waste treatment.

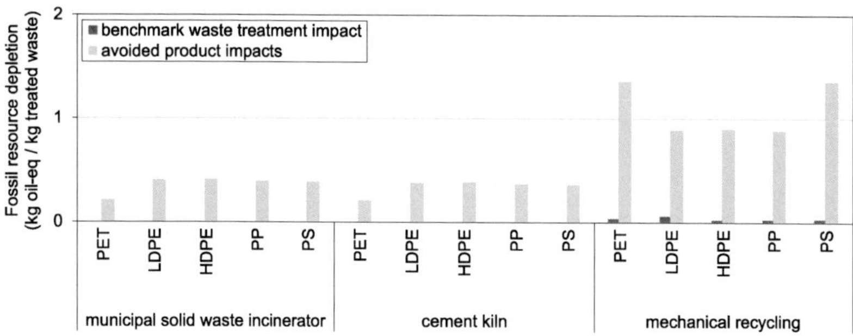

Figure A.4: The direct fossil resource depletion of benchmark waste treatments (dark grey) and the credit avoided products (light grey) are shown. All plastic packaging wastes are shown for each benchmark waste treatment.

Energy recovery in municipal solid waste incinerators

The global warming impact of energy recovery in municipal solid waste incinerators results from direct emissions of plastic packaging waste incineration. Global warming impacts range from 2.01 kg CO_2-eq for PET to 2.85 kg CO_2-eq for PS. The difference is due to varying carbon contents of each plastic packaging waste, i.e. 63 wt. % for PET and 92 wt. % for PS.

Avoided global warming impacts for products are based on avoided average European grid electricity and district heating production. Nowadays, the European electricity mix includes a share of renewable energies of about 30 % (Eurostat, 2018). District heating is based on 44 % (Connolly et al., 2014) natural gas boilers which have higher efficiencies than municipal solid waste incinerators. Thus, avoided global warming impacts are lower than direct global warming impacts of incineration itself. As a result, treating plastic packaging waste in municipal solid waste incinerators is disadvantageous from the perspective of global warming impacts.

In case of fossil resource depletion, direct impacts of municipal solid waste incinerators are negligible. However, moderate credits are achieved from substituting electricity production and district heating.

Energy recovery in cement kilns

The global warming impacts of energy recovery in cement kilns mainly are due to direct emissions of plastic packaging waste incineration and, thus, are similar to those of the energy recovery in solid waste incinerators. Environmental impacts are avoided based on the substitution of lignite as energy source.

Lignite results in high specific global warming impacts of about 0.1 kg CO_2-eq per MJ thermal energy. These high specific global warming impacts originate from the carbon content of approx. 64.27 wt. % of lignite and the relatively low net calorific value of 23.64 MJ/kg (Ecoinvent, 2020) (dry mass basis).

In comparison, polyolefins (HDPE, LDPE, PP) and PS offer higher net calorific values ranging from 41.96 MJ/kg for PS to 44.60 MJ/kg for HDPE. PET (20.48 MJ/kg) is characterized by net calorific values slightly lower than lignite, because of its increased oxygen content. Furthermore, polyolefins and PS have higher carbon contents ranging from 85.63 wt. % for polyolefins to 92.30 wt. % for PS. The carbon content of PET (62.50 wt. %) is, again, in the range of lignite.

Based on the carbon contents and net calorific values, energy recovery of plastic packaging wastes result in specific global warming impacts ranging from 0.070 kg CO_2-

eq per MJ for polyolefins and 0.087 kg CO_2-eq per MJ for PET. Thus, the credit for lignite substitution (0.1 kg CO_2-eq per MJ) is higher than direct global warming impacts from incinerating polyolefins (0.070 kg CO_2-eq per MJ) and PET (0.087 kg CO_2-eq per MJ).

The fossil resource depletion of energy recovery in cement plants is negligible. All plastic packaging waste result in positive credits for fossil resource depletion. However, the fossil resource depletion credit of PET is nearly half of the value derived for polyolefins and PS, because of the lower net calorific value of PET.

Mechanical recycling

The environmental impacts of mechanical recycling include electricity production for wet-processing and extrusion as well as the treatment of residual wastes. Avoided environmental impacts are based on the production of virgin polymers and the substitution factor to account for downcycling.

Environmental impacts of mechanical recycling are in a narrow range for all plastic packaging wastes. The global warming impacts range from 0.41 kg CO_2-eq for HDPE to 0.48 kg CO_2-eq for LDPE. The differences are due to changes in the amount of residual plastic waste that is treated by energy recovery and the required electricity for wet processing and extrusion. Mechanical recycling of LDPE has the highest global warming impacts since it combines the lowest efficiency and highest electricity demand.

The credit for virgin polymer production is largely influenced by the type of virgin polymer as well as the substitution factor. Interestingly, polyolefins are disadvantageous with regard to the avoided global warming impacts and the substitution factor: the credit for global warming impacts is up to 59 % lower than for PET and PS. In case of fossil resource depletion, the credit is reduced by up to 33 % for polyolefins in comparison to PET and PS. The difference occurs because the production of polyolefins is less energy-intensive than the production of virgin PET and PS. Thus, the credit for virgin polyolefins is smaller than for PET and PS. Additionally, polyolefins suffer from increased downcycling and, thus, have lower substitution factors of 0.7 compared to 1 for PET and 0.9 for PS.

A.7 Modeling of chemical recycling

Chemical recycling is modeled according to the idealized case (cf. Section 3.1.2). Environmental impacts associated with the production of reactants, residual waste treatment and auxiliary supply based on data from ecoinvent (Ecoinvent, 2020) are included, if not stated otherwise. Hydrogen and natural gas production is based on data provided by thinkstep (Sphera, 2019).

(1) **Refinery feedstock:** For polyolefins and PS, it is assumed that the plastic packaging waste directly substitutes refinery feedstock, e.g. petroleum oil (direct utilization) without any chemical reaction. Furthermore, a substitution on an equivalent mass basis is assumed. The assumption is based on the small difference of carbon and hydrogen contents of polyolefins (C: 85.63 wt. %, H: 14.37 wt. %) and polystyrene (C: 92.26 wt. %, H: 7.74 wt. %), compared to oil (C: 85 wt. %, H: 12 wt. %) (Speight, 2014). Due to the similar elementary compositions, refinery units are expected to produce the same range of products if polyolefins and PS are used to partly substitute petroleum oil (Lopez et al., 2017; Barbarias et al., 2018).

(2) **Fuel production:** Gaseous fuel products are assumed to substitute natural gas (Honus et al., 2018b). Liquid fuel products are assumed to substitute gasoline or diesel (Kalargaris et al., 2017; Anuar Sharuddin et al., 2018). However, the composition and net calorific values of gaseous and liquid fuel products may differ from their substitutes, natural gas, gasoline and diesel. Therefore, the system boundaries for evaluation need to be extended by the subsequent incineration and energy generation to account for the difference of composition and net calorific values. Equal energy efficiencies are assumed for both, plastic-derived fuel products and their benchmark fuels.

To calculate the minimal energy input (Q_H), net calorific values from literature are used (Table A.7) (Linstrom, 2018). Gaseous fuel products have lower net calorific values compared to natural gas (50.4 MJ/kg) (Ecoinvent, 2020). Depending on the plastic packaging waste, heating values of gaseous fuel products range between 8.19 MJ/kg (PET derived, omitted in analysis, cf. Section 3.1.5.2) and 48.7 MJ/kg (PS derived).

For liquid fuel products, the heating values range between 28.2 MJ/kg (PET derived), 40 MJ/kg (PE derived), 40.8 MJ/kg (PP derived) and 43 MJ/kg (PS derived). Except for PET, the derived values are in the same range as for gasoline and diesel (approx. 43 MJ/kg) (Anuar Sharuddin et al., 2016).

For the modeling of fuel production, no external energy source is used to supply process heat for depolymerization, because the energy demand for gasification or pyrolysis is assumed to be supplied internally by the produced fuels. A detailed list of composition and heating values of each polymer and the substituted fuel can be found in Table A.6. Furthermore, all carbon in plastic-derived fuels and substituted fuels is assumed to be converted to CO_2.

(3) **Monomer production:** Conventional chemical production is substituted. However, in case of dimethyl terephthalate production, no consistent life cycle assessment dataset could be obtained from life cycle assessment databases. To calculate environmental impacts, life cycle inventory datasets for the production of dimethyl terephthalate are based on data provided in technical process reports and used aggregated datasets from ecoinvent for feedstock and energy carrier emissions (IHS Markit, 2018).

Dimethyl terephthalate is produced from para-xylene. The process includes four process steps: oxidation, esterification, separation and purification, and catalyst recovery. The derived global warming impact per kg dimethyl terephthalate sums up to 2.42 kg CO_2-eq emission. Fossil resource depletion per kg dimethyl terephthalate equals 1.56 kg oil-eq.

To calculate the minimal energy input (Q_H) for monomer production, standard enthalpies of formation (Afeefy et al., 2018; Burgess, 2018; Glushko Thermocenter, 2018) are used, based on reported reaction enthalpies (Roberts, 1950) of polymers. Energy is supplied from a natural gas boiler.

(4) **Chemical upcycling:** In case of PET, the substitution of cyclohexane dimethanol and ethylene glycol is assumed. However, no data about global warming impacts or fossil resource depletion could be obtained from life cycle assessment databases for cyclohexane dimethanol. To calculate environmental impacts, cyclohexane dimethanol is modeled based on data provided in technical process reports (IHS Markit, 2018).

The production of cyclohexane dimethanol is modeled based on the hydrogenation of dimethyl terephthalate in four fixed-bed reactors. Subsequently, cyclohexane dimethanol is purified by extracting methanol that is formed as a by-product. By-product methanol is credited by its conventional production from synthesis gas. The derived global warming impact per kg cyclohexane dimethanol sums up to 5.15 kg CO_2-eq emission. Fossil resource depletion is 2.66 kg oil-eq. To calculate the minimal energy input (Q_H), the standard enthalpies of formation are used (Afeefy et al., 2018; Burgess, 2018; Glushko Thermocenter, 2018; Thomson, 1996). Energy is supplied from natural gas boiler.

A.8 Calculating the reaction enthalpy

The minimal energy requirement Q_H can be calculated by standard thermodynamic relations.

Equation A.1 uses the standard enthalpies of formation of products $\Delta h^0_{f,j}$, reactants $\Delta h^0_{f,i}$ and the polymer $\Delta h^0_{f,p}$:

$$Q_H = \sum_j m_j \Delta h^0_{f,j} - \sum_i m_i \Delta h^0_{f,i} - m_p \Delta h^0_{f,p} \tag{A.1}$$

Based on Hess's law of constant heat summation, equation A.2 calculates the minimal energy requirement Q_H based on the gross calorific values of products $\Delta h^0_{c,j}$, reactants $\Delta h^0_{c,i}$ and the polymer $\Delta h^0_{c,p}$ if water is assumed in liquid state:

$$Q_H = \sum_j m_j \Delta h^0_{c,j} - \sum_i m_i \Delta h^0_{c,i} - m_p \Delta h^0_{c,p} \tag{A.2}$$

A.9 Datasets for chemical recycling

Table A.6: Composition and net calorific values of polymers and substituted fuels.

elementary composition in wt. %[1]	PET	HD/ LDPE	PP	PS	gasoline	diesel	natural gas[2]
C	62.50	85.63	85.63	92.26	85.50	86.50	72.93
H	4.20	14.37	14.37	7.74	14.40	13.20	23.98
N							2.81
O	33.30						0.28
S					0.10	0.30	
net calorific value in MJ/kg	23.22[3]	44.60[3]	42.66[3]	41.96[3]	43.74[1]	42.82[1]	50.50

[1] Phyllis2 (2019)
[2] Ecoinvent (2020)
[3] Walters et al. (2000)

Table A.7: Net calorific values of fuel products from plastic packaging wastes.

polymer	net calorific value of gaseous fuels in MJ/kg [1]	net calorific value of liquid fuels in MJ/kg[2]
PET	8.19	28.20
PP	46.12	40.80
HD LDPE	46.36	40.00
PS	48.67	43.00

[1] Honus et al. (2018a,b)
[2] Anuar Sharuddin et al. (2016)

Table A.8: Chemical recycling datasets for refinery feedstock substitution.

Type	plastic packaging waste	unit	HDPE	LDPE	PP	PET	PS
outputs	refinery feedstock	kg/kg	0.846	0.828	0.846	0.846	-
product	waste water to treatment	kg/kg	0.100	0.100	0.100	0.100	0.100
	waste residues to energy recovery	kg/kg	0.054	0.072	0.054	0.054	0.018

Table A.9: Chemical recycling datasets for gaseous or liquid fuel production. Only the net energy output is shown. In both cases either gaseous fuels or liquid fuels are produced. Energy demand of pyrolysis or gasification is supplied internally by the utilization of produced fuels.

Type	plastic packaging waste	unit	HDPE	LDPE	PP	PET	PS
outputs	gaseous fuel	MJ/kg	37.22	36.43	36.52	33.71	-
	liquid fuel	MJ/kg	37.22	36.43	36.52	33.71	19.19
product	waste water to treatment	kg/kg	0.100	0.100	0.100	0.100	0.100
	waste residues to energy recovery	kg/kg	0.054	0.072	0.054	0.054	0.018

Table A.10: Chemical recycling datasets for monomer production and chemical upcycling from plastic packaging wastes. The production of terephthalic acid is denoted PET TPA, while the production of dimethyl terephthalate is denoted PET DMT. The chemical upcycling of PET is denoted PET up.

Type	**plastic packaging waste**	**unit**	**HD PE**	**LD PE**	**PP**	**PET TPA**	**PS**	**PET DMT**	**PET up**
inputs	methanol	kg/kg						0.294	
	heat	MJ/kg	2.817	2.757	2.352	0.120	0.609	0.180	3.41
	water	kg/kg				0.165			
	hydrogen	kg/kg							0.06
outputs	ethylene	kg/kg	0.846	0.828					
	propylene	kg/kg			0.846				
	styrene	kg/kg					0.846		
	dimethyl terephthalate	kg/kg						0.891	
	terephthalic acid	kg/kg				0.763			
	ethylene glycol	kg/kg				0.285		0.285	0.285
	cyclohexane dimethanol	kg/kg							0.662
product	waste water to treatment	kg/kg	0.100	0.100	0.100	0.100	0.100	0.100	0.100
	waste residues to energy recovery	kg/kg	0.054	0.072	0.054	0.018	0.054	0.018	0.018

Table A.11: Used standard enthalpies of formation and molar masses from the NIST Chemistry Webbook (Afeefy et al., 2018; Burgess, 2018) and DIPPR (Thomson, 1996).

substance	$\Delta h^0_{f,i}$ **[kJ/mol]**	M_i **[g/mol]**	$\Delta h^0_{f,i}$ **[MJ/kg]**
water (gaseous)	-241.83	18.02	-13.42
carbon dioxide (gaseous)	-393.51	44.01	-8.94
ethylene (gaseous)	52.40	28.05	1.87
propylene (gaseous)	20.41	42.08	0.49
terephthalic acid (solid)	-816.90	166.13	-4.92
ethylene glycol (liquid)	-460.00	62.07	-7.41
cyclohexanedimethanol (solid)	469.70	144.21	3.26
methanol (liquid)	-238.40	32.04	-7.44
dimethyl terephthalate (solid)	-710.00	194.18	-3.66
water (liquid)	-285.30	18.02	-15.84
styrene (liquid)	103.40	104.15	0.99

Table A.12: Chemical recycling datasets for gaseous or liquid fuel production. Only the net energy output is shown. In both cases either gaseous fuels or liquid fuels are produced.

polymer type	**unit**	**HDPE**	**LDPE**	**PP**	**PET**	**PS**
raw materials	MJ/kg	1.868	1.868	0.485	-6.645	0.993
reaction enthalpy[1]	MJ/kg	-3.330	-3.330	-2.780	-0.136	-0.720
water as by-product	MJ/kg				-2.969	
enthalpy of formation	MJ/kg	-1.462	-1.462	-2.295	-3.811	0.273

[1] Roberts (1950)

A.10 Life cycle assessment results for chemical recycling

The results for the global warming impact of chemical recycling pathways are shown in Figure A.5. Fossil resource depletion is shown in Figure A.6. Results are shown for each category of chemical recycling: (i) refinery feedstock, (ii) fuel production, (iii) monomer production and (iv) chemical upcycling.

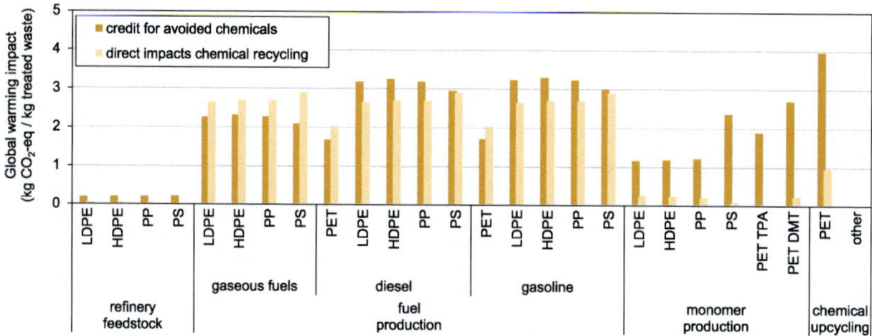

Figure A.5: Direct global warming impacts of chemical recycling (light orange) and credit for avoided chemicals (dark orange). Chemical recycling technologies are shown for 4 categories: (i) refinery feedstock, (ii) fuel production, (iii) monomer production and (iv) chemical upcycling. For each category, plastic packaging wastes are shown that can be used to produce the subsequent chemical product. In case of PET, two monomer production pathways are shown: terephthalic acid (PET TPA) and dimethyl terephthalate (PET DMT).

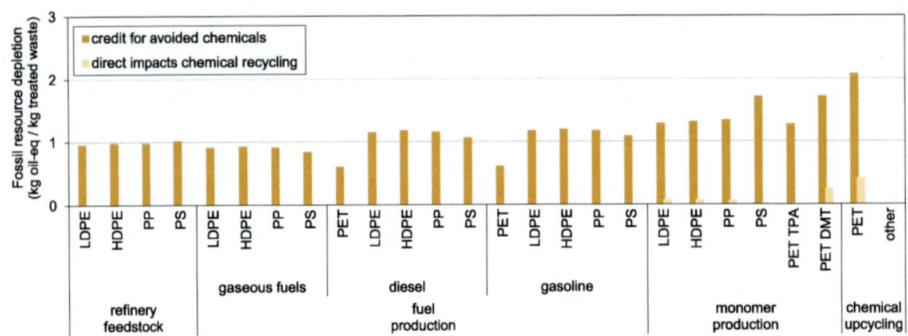

Figure A.6: Direct fossil resource depletion of chemical recycling (light orange) and credit for avoided chemicals (dark orange). Chemical recycling technologies are shown for 4 categories: (i) refinery feedstock, (ii) fuel production, (iii) monomer production and (iv) chemical upcycling. For each category, plastic packaging wastes are shown that can be used to produce the subsequent chemical product. In case of PET, two monomer production pathways are shown: terephthalic acid (PET TPA) and dimethyl terephthalate (PET DMT).

Refinery feedstock

Utilizing plastic packaging waste as refinery feedstock leads to negligible direct environmental impacts, because sorted plastic packaging waste is assumed to substitute petroleum oil without any further processing. Thus, each kg of polymer in the plastic packaging waste substitutes 1 kg of crude petroleum oil. The credit for 1 kg of petroleum oil includes exploration, drilling and extraction as well as long- and short-distance distribution (Ecoinvent, 2020). Still, the global warming impact of one kilogram of crude petroleum oil is relatively low (Figure A.5, left side, dark orange).

For fossil resource depletion, the credit for petroleum oil is higher (Figure A.6, left side, dark orange). Thus, chemical recycling to refinery feedstock seems promising solely from the perspective of fossil resource depletion.

Fuel production - gaseous fuels

The chemical recycling to gaseous fuels comprises the environmental impacts of (i) residual waste treatment, (ii) thermal energy supply as well as (iii) the subsequent incineration to account for differences of net calorific values and carbon content of plastic-derived and substituted fuels. Credited environmental impacts are based on an equivalent amount of thermal energy from natural gas based on equal net calorific values.

For all plastic packaging waste, direct global warming impacts of chemical recycling are higher than the credit for gaseous fuel products. The results can be explained by the inherent properties of plastic-derived gaseous fuels: The net calorific values of gaseous fuels are between 3 % (PS-derived) and 9 % (PP-derived) lower than the net calorific value of natural gas. At the same time, carbon contents of gaseous fuels are between 27 % (PS-derived) and 17 % (PP-derived) higher than natural gas. Therefore, the specific global warming impact per mega joule is higher for plastic-derived gaseous fuels than for natural gas, making natural gas the favorable source for thermal energy in terms of global warming impact.

However, the production of gaseous fuels saves significant amounts of fossil resources: between 0.93 kg oil-eq (HDPE derived) and 0.84 kg oil-eq (PS derived) can be avoided by producing gaseous fuels. The slightly lower value for PS can be explained by the higher thermal energy consumption calculated from the difference of net calorific value of PS (41.96 MJ/kg (Walters et al., 2000)) and derived fuel (48.67 MJ/kg (Honus et al., 2018a)).

In conclusion, the production of gaseous fuels seems promising only in terms of fossil resource depletion. Using plastic packaging waste for gaseous fuels increases greenhouse gas emissions compared to the utilization of natural gas und thus seems disadvantageous from the perspective of global warming impacts.

Fuel production - liquid fuels

Like gaseous fuels, environmental impacts for liquid fuel production comprises (i) residual waste treatment, (ii) thermal energy supply as well as (iii) the subsequent incineration of plastic-derived liquid fuels. The credit of environmental impacts is based on the substitution of conventional diesel and gasoline production based on equal net calorific values. For liquid fuel products, results vary qualitatively between PET and the other plastic packaging wastes.

For liquid fuel production from PET, direct global warming impacts of chemical recycling are higher than credits from gasoline and diesel production. PET-derived liquid fuels have lower net calorific values than gasoline (36 %) and diesel (35 %) due to increased inherent oxygen content in PET. Also, the credit in terms of fossil resource depletion is lower compared to other plastic packaging wastes.

In contrast to PET, net calorific values of liquid fuels derived from polyolefins and PS are in the same range as gasoline and diesel. Therefore, polyolefins and PS have a higher credit than the direct environmental impacts.

In summary, liquid fuel products for polyolefins and PS seems promising with regard to both, global warming impacts and fossil resource depletion, while PET-derived liquid fuels solely offer promising results for fossil resource depletion.

Monomer production

The environmental impact of chemical recycling for monomer production includes (i) the treatment of residual wastes, (ii) the production of reactants as well as (iii) thermal energy supply. Credits are based on the environmental impacts of conventional production of monomers. The results for monomer production are qualitatively similar for both the global warming impact and fossil resource depletion and, thus, are discussed together. For all plastic packaging wastes, direct environmental impacts of chemical recycling are smaller than the credit for conventional monomer production.

Interestingly, avoided environmental impacts for PET and PS are higher than for polyolefins. The higher values result from higher avoided environmental impacts of terephthalic acid, dimethyl terephthalate (both PET) and styrene (PS) production.

If terephthalic acid is produced from PET, only water is required as reactant (Paszun and Spychaj, 1997). Dimethyl terephthalate production requires methanol as reactant (Paszun and Spychaj, 1997) and, thus, increases environmental impacts of chemical recycling. However, avoided environmental impacts increase if dimethyl terephthalate is produced, because conventional dimethyl terephthalate is produced from terephthalic acid and methanol by esterification (Elvers and Ullmann, 2011). Esterification of terephthalic acid forms water as by-product that has to be separated from dimethyl terephthalate (Elvers and Ullmann, 2011), resulting in increased environmental impacts from energy supply. The production of methanol is avoided if dimethyl terephthalate is produced from PET packaging waste, offering an additional advantage for chemical recycling (Paszun and Spychaj, 1997).

For all polyolefins, environmental impacts of chemical recycling and avoided environmental impacts are similar. PET and PS, in contrast, have higher avoided environmental impacts from monomer substitution. Thus, monomer production of PET and PS seems more promising than monomer production from polyolefins.

Chemical upcycling

In case of chemical upcycling, the environmental impacts include (i) the treatment of residual wastes, (ii) the production of reactants as well as (iii) thermal energy supply. For chemical upcycling, only the production of cyclohexane dimethanol and ethylene glycol from PET packaging waste is assessed.

For chemical upcycling of PET to cyclohexane dimethanol, hydrogen is used as reactant. Hydrogen is produced from steam methane reforming and, thus, results in high specific emission of approx. 10 kg CO_2-eq emissions per kg of hydrogen. Thus, direct environmental impacts are high for the chemical upcycling of PET.

However, the credit given for the conventional cyclohexane dimethanol production is also high (Figure A.5, right side, dark orange). The high credit can be explained based on the conventional production route of cyclohexane dimethanol: the conventional production route is based on the hydrogenation of dimethyl terephthalate with pure hydrogen. Furthermore, methanol is formed during hydrogenation as a by-product and has to be separated from the product stream. Thus, the conventional production of cyclohexane dimethanol has high global warming impacts by supplying pure hydrogen and additional energy for product separation. In summary, chemical upcycling of PET to cyclohexane dimethanol represents a promising candidate for chemical recycling in terms of the global warming impact and fossil resource depletion.

A.11 Comparison to cement kilns (biomass or natural gas as fuel)

negative potential	positive potential	Global warming impact [kg CO_2-eq]						Fossil resource depletion [kg oil eq]					
		refinery feedstock	gaseous fuels	gasoline	diesel	monomers	chemical upcycling	refinery feedstock	gaseous fuels	gasoline	diesel	monomers	chemical upcycling
(2) energy recovery in cement kiln (natural gas)	PET TPA			0.38	0.35	2.46	3.77			0.13	0.11	0.75	1.19
	PET DMT					3.20						0.99	
	LDPE	0.39	-0.07	0.83	0.77	1.15		0.07	-0.01	0.28	0.26	0.31	
	HDPE	0.41	-0.06	0.85	0.80	1.19		0.07	-0.01	0.29	0.27	0.32	
	PP	0.51	0.00	0.90	0.84	1.36		0.11	0.01	0.31	0.28	0.40	
	PS	0.79	-0.14	0.69	0.63	2.78		0.17	-0.04	0.23	0.21	0.83	
(2) energy recovery in cement kiln (biomass)	PET TPA			1.67	1.64	3.76	5.06			0.62	0.61	1.24	1.68
	PET DMT					4.49						1.48	
	LDPE	2.72	2.26	3.15	3.10	3.48		0.96	0.87	1.17	1.15	1.20	
	HDPE	2.79	2.32	3.23	3.18	3.57		0.98	0.89	1.20	1.17	1.22	
	PP	2.79	2.27	3.17	3.11	3.64		0.98	0.88	1.17	1.15	1.27	
	PS	3.02	2.10	2.92	2.87	5.02		1.02	0.81	1.08	1.06	1.68	

Figure A.7: Environmental potential for chemical recycling compared to energy recovery in cement kilns that uses natural gas or biomass as fuel. Red indicates a negative environmental potential. Green indicates a positive environmental potential. White indicates environmental potentials close to zero. Grey indicates that chemical recycling does not exist or has been omitted. PET can be used to produce ethylene glycol and two types of monomers: terephthalic acid denoted PET TPA) and dimethyl terephthalate (denoted PET DMT).

A.12 Additional results of sensitivity analysis

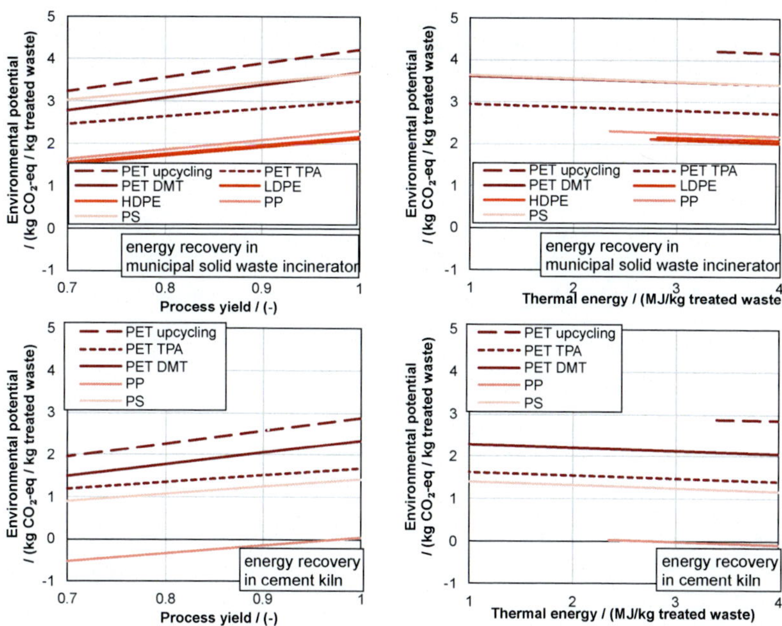

Figure A.8: Environmental potential in terms of the global warming impacts for monomer production and chemical upcycling in comparison to energy recovery in municipal waste incinerators and cement kilns. The marked values (x) represent the highest reported values of conversion rates for monomer production by chemical recycling. Terephthalic acid production from PET is denoted PET TPA, while dimethyl terephthalate production is denoted PET DMT.

A.13 Datasets used from life cycle assessment databases

Table A.13: Sources of life cycle assessment datasets used in the case-study.

process	**Source**
natural gas production hydrogen production	Gabi database (Sphera, 2019)
market group for heat, central or small-scale, natural gas market for petrol, unleaded market group for diesel, low-sulfur market for lignite market for electricity electricity production, wind, <1MW turbine, onshore market for petroleum treatment of wastewater, average, capacity 1E9l/year polyethylene production, high density, granulate polyethylene production, low density, granulate polyethylene terephthalate production, granulate, amorphous polystyrene production, general purpose polypropylene production market group for heat, central or small-scale, other than natural gas ethylene production, average ethylene glycol production xylene production market group for tap water propylene production styrene production purified terephthalic acid production market for methanol steam production, in chemical industry market for sodium hydroxide, without water, in 50 % solution state	ecoinvent (Ecoinvent, 2020)

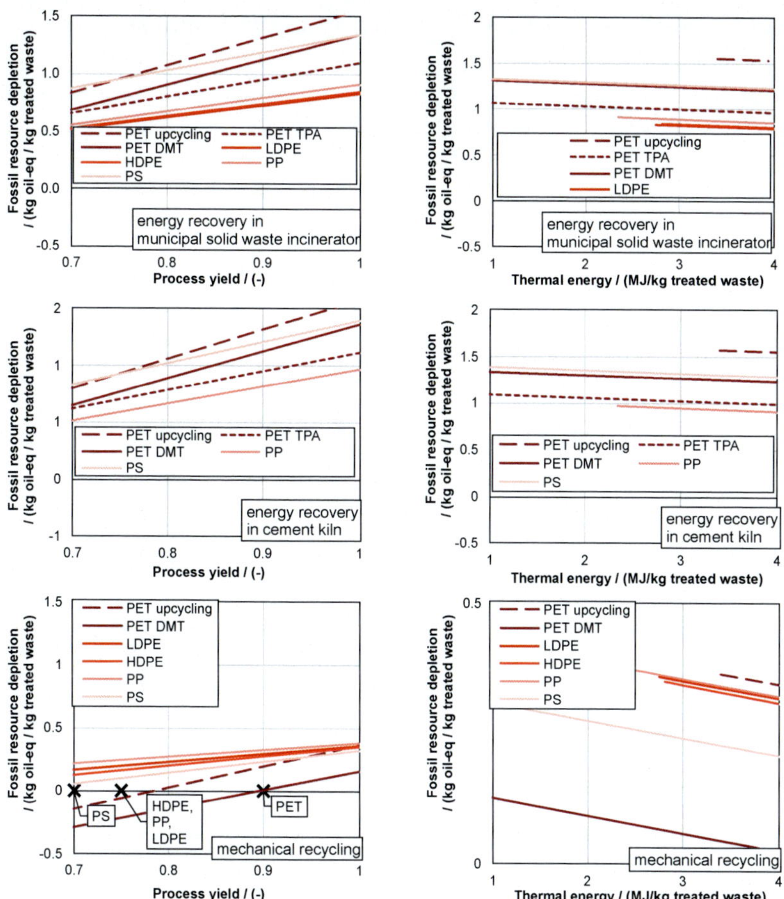

Figure A.9: Environmental potential are shown depended on the process yield (left) and thermal energy demand (right) of chemical recycling. Monomer production and chemical upcycling are compared to all three benchmark waste treatment technologies. The values for the environmental potential (y-axis). The marked values (x) represent the highest reported values of conversion rates for monomer production by chemical recycling. Terephthalic acid production from PET is denoted PET TPA, while dimethyl terephthalate production is denoted PET DMT.

A.14 Tabulated values of environmental impacts

type	unit	HDPE	LDPE	PP	PET	PS	PET II
		refinery feedstock					
impact benchmark waste treatment	kg CO_2-eq	0.03	0.04	0.03		0.01	
avoided product impacts	kg CO_2-eq	0.19	0.18	0.19		0.2	
impact benchmark waste treatment	kg oil-eq	0	0	0		0	
avoided products impacts	kg oil-eq	0.95	0.93	0.95		0.99	
		gaseous fuels					
impact benchmark waste treatment	kg CO_2-eq	2.67	2.63	2.67		2.88	
avoided product impacts	kg CO_2-eq	2.36	2.31	2.32		2.14	
impact benchmark waste treatment	kg oil-eq	0	0	0		0	
avoided products impacts	kg oil-eq	0.9	0.88	0.88		0.81	
		liquid fuels (gasoline)					
impact benchmark waste treatment	kg CO_2-eq	2.67	2.63	2.67	2.02	2.88	
avoided product impacts	kg CO_2-eq	3.17	3.1	3.11	1.63	2.87	
impact benchmark waste treatment	kg oil-eq	0	0	0	0	0	

continued on next page

Table A.14 – *continued from previous page*

type	unit	HDPE	LDPE	PP	PET	PS	PET II
avoided products impacts	kg oil-eq	1.05	1.03	1.03	0.54	0.95	
		liquid fuels (diesel)					
impact benchmark waste treatment	kg CO_2-eq	2.67	2.63	2.67	2.02	2.88	
avoided product impacts	kg CO_2-eq	3.22	3.15	3.16	1.66	2.92	
impact benchmark waste treatment	kg oil-eq	0	0	0	0	0	
avoided products impacts	kg oil-eq	1.12	1.1	1.1	0.58	1.02	
		monomer production					
impact benchmark waste treatment	kg CO_2-eq	0.23	0.24	0.2	0.02	0.07	0.21
avoided product impacts	kg CO_2-eq	1.18	1.16	1.21	1.74	2.22	2.6
impact benchmark waste treatment	kg oil-eq	0.1	0.09	0.08	0	0.02	0.3
avoided products impacts	kg oil-eq	1.37	1.34	1.4	1.32	1.76	1.86
		chemical upcycling					
impact benchmark waste treatment	kg CO_2-eq				0.938		
avoided product impacts	kg CO_2-eq				3.867		

continued on next page

Table A.14 – *continued from previous page*

type	unit	HDPE	LDPE	PP	PET	PS	PET II
impact benchmark waste treatment	kg oil-eq				0.402		
avoided products impacts	kg oil-eq				2.172		
energy recovery municipal solid waste incinerator							
impact benchmark waste treatment	kg CO_2-eq	2.64	2.59	2.64	2.01	2.85	2.01
avoided product impacts	kg CO_2-eq	1.43	1.39	1.36	0.73	1.33	0.73
impact benchmark waste treatment	kg oil-eq	0	0	0	0	0	0
avoided products impacts	kg oil-eq	0.43	0.42	0.41	0.22	0.4	0.22
energy recovery in cement kilns							
impact benchmark waste treatment	kg CO_2-eq	2.64	2.59	2.64	2.01	2.85	2.01
avoided product impacts	kg CO_2-eq	3.8	3.72	3.63	2.06	3.57	2.06
impact benchmark waste treatment	kg oil-eq	0	0	0	0	0	0
avoided products impacts	kg oil-eq	0.37	0.36	0.35	0.2	0.35	0.2
mechanical recycling							
impact benchmark waste treatment	kg CO_2-eq	0.41	0.48	0.41	0.45	0.44	

continued on next page

Table A.14 – *continued from previous page*

type	unit	HDPE	LDPE	PP	PET	PS	PET II
avoided product impacts	kg CO_2-eq	0.98	1.01	0.97	2.25	2.38	
impact benchmark waste treatment	kg oil-eq	0.02	0.06	0.02	0.04	0.02	
avoided products impacts	kg oil-eq	0.95	0.94	0.96	1.44	1.44	

A.15 Terrestrial acidification and eutrophication potentials

Besides global warming and fossil resource depletion, the environmental impacts most commonly assessed for plastic waste management are terrestrial acidification and marine/freshwater eutrophication (Lazarevic et al., 2010). Terrestrial acidification is caused by emissions of nitrogen and sulfur, e.g. nitrogen oxides, ammonia and sulfur dioxide, into soil or freshwater (Goedkoop et al., 2013; Hauschild and Huijbregts, 2015). In particular, the combustion of fossil fuels or agricultural processes largely contributes to terrestrial acidification. Freshwater eutrophication is increased due to phosphorous emissions such as phosphate, while marine eutrophication is increased by nitrous emissions, e.g. ammonia, nitrate or nitrous oxides. Both eutrophication impacts are dominated by agricultural and farming processes but also industrial point sources such as waste water treatment plants (Goedkoop et al., 2013; Hauschild and Huijbregts, 2015).

Like for global warming and fossil resource depletion, chemical recycling routes with positive and negative environmental potentials for acidification and eutrophication are identified. Thus, some chemical recycling routes increase terrestrial acidification, marine and freshwater eutrophication compared to municipal waste incinerators, even assuming ideal conditions. At the same time, other routes have the potential to reduce these impacts. With respect to the specific routes, the findings for acidification and eutrophication differ from the results obtained for global warming impacts and fossil resource depletion indicating potential trade-offs between environmental impacts.

Chemical recycling vs. energy recovery in municipal solid waste incin-

erators

Recycling plastic packaging to produce refinery feedstock and fuels has no potential to reduce environmental impacts regarding neither marine and freshwater eutrophication, nor terrestrial acidification if compared to energy recovery in municipal solid waste incinerators (cf. Figure A.10 to Figure A.12). The increase in eutrophication and acidification must be weighed against the potential to reduce global warming impacts and fossil resource depletion. Monomer production from polyolefins also has no environmental potential compared to municipal solid waste incinerators: Municipal solid waste incinerators have high credits in term of acidification and eutrophication due to the substitution of conventional electricity and heat production, e.g. the substitution of fossil-based energy supply as major source for both impact categories (Hauschild and Huijbregts, 2015).

In contrast, monomer production and chemical upcycling of PET and PS have the possibility to reduce eutrophication and acidification. Monomer production and chemical upcycling have a large credit from the substitution of conventional monomer and cyclohexane di-methanol production.

Thus, the analysis of acidification and eutrophication impacts also shows that it is necessary to produce value-added chemicals from plastic packaging waste, such as monomers or upcycled chemicals, in order to reduce environmental impacts compared to energy recovery in municipal solid waste incinerators.

Chemical recycling vs. energy recovery in cement kilns

The comparison between chemical recycling and energy recovery in cement kilns differs for terrestrial acidification from the findings for global warming impacts. Chemical recycling offers the potential to reduce terrestrial acidification in all cases, because chemical recycling avoids the emission of inorganic salts used during oil drilling as well as SO_2 and NOx emissions during refinery operations. Both, drilling for oil and refinery operations, largely contribute to terrestrial acidification of chemical production (Althaus et al., 2007). These results are in-line with the result for fossil resource depletion, because most of the environmental impacts result from oil and refinery operations in both cases.

For marine eutrophication, results for the environmental potential are mostly in line with the results obtained for global warming impacts: Only if chemical recycling produces monomers or cyclohexane di-methanol from PET, eutrophication can potentially be reduced compared to energy recovery in cement kilns. Plastic packaging wastes substitute coal in the cement kilns and, thereby, avoids phosphorous emissions and wastewater treatment during lignite mining. Mining has particularly high impacts on freshwater eutrophication such that even monomer production and chemical upcycling have no potential to reduce environmental impacts.

Chemical recycling vs. mechanical recycling

For acidification as well as eutrophication results differ from those for global warming and fossil resource depletion if chemical recycling is compared to mechanical recycling. Using plastic packing waste for refinery feedstock, gaseous fuel and, in the case of PET, for monomer production compared to mechanical recycling has no environmental potential in terms of terrestrial acidification. The negative potentials are based on the avoided terrestrial acidification impacts from the production of virgin polymers and the corresponding process energy demands. In contrast, chemical recycling has the potential to reduce terrestrial acidification for conversion of polyolefins to refinery feedstock, liquid fuels and monomers, because of the downcycling of polyolefins during mechanical recycling and the resulting lower credit for the substitution of virgin polymer production.

In case of freshwater and marine eutrophication, results are similar to the result obtained for fossil resource depletion: for polyolefins, all chemical recycling technologies have the potential to reduce environmental impacts with regard to freshwater and marine eutrophication. In contrast to polyolefins, PET and PS have no potential to reduce freshwater and marine eutrophication if refinery feedstock or fuels are produced by chemical recycling. Mechanical recycling receives a high credit for freshwater and marine eutrophication by substituting virgin PET and PS. The high credit for substituting virgin PET production could be explained by the polycondensation reaction and the required wastewater treatment, a major source for eutrophication during chemical production. Chemical upcycling has the potential to reduce freshwater and marine eutrophication.

A.16 Environmental potentials for acidification and eutrophication

negative potential	positive potential	Terrestrial acidification [kg SO2 eq]					
		refinery feedstock	gaseous fuels	gasoline	diesel	monomers	chemical upcycling
(1) energy recovery in waste incinerator	PET TPA			-5.29E-04	-9.01E-04	3.00E-03	4.73E-03
	PET DMT					3.76E-03	
	LDPE	-3.70E-03	-4.67E-03	-9.53E-04	-1.66E-03	-4.60E-03	
	HDPE	-3.78E-03	-4.77E-03	-9.75E-04	-1.70E-03	-4.70E-03	
	PP	-3.50E-03	-4.52E-03	-7.93E-04	-1.50E-03	-4.10E-03	
	PS	-3.29E-03	-4.52E-03	-1.08E-03	-1.73E-03	2.31E-03	
(2) energy recovery in cement kiln (natural gas)	PET TPA			2.49E-03	2.12E-03	6.02E-03	7.75E-03
	PET DMT					6.78E-03	
	LDPE	1.98E-03	1.01E-03	4.73E-03	4.03E-03	1.08E-03	
	HDPE	2.03E-03	1.04E-03	4.84E-03	4.12E-03	1.11E-03	
	PP	2.04E-03	1.03E-03	4.75E-03	4.05E-03	1.45E-03	
	PS	2.16E-03	9.29E-04	4.37E-03	3.72E-03	7.77E-03	
(3) mechanical recycling	PET TPA			-5.95E-03	-6.32E-03	-2.42E-03	-6.84E-04
	PET DMT					-1.66E-03	
	LDPE	-1.79E-03	-2.76E-03	9.59E-04	2.53E-04	-2.69E-03	
	HDPE	-1.26E-03	-2.26E-03	1.54E-03	8.23E-04	-2.18E-03	
	PP	-1.09E-03	-2.11E-03	1.62E-03	9.12E-04	-1.69E-03	
	PS	-5.01E-03	-6.24E-03	-2.80E-03	-3.45E-03	5.94E-04	

Figure A.10: Environmental potential in terms of terrestrial acidification for chemical recycling compared to energy recovery in cement kilns that uses lignite as fuel as well as energy recovery in municipal waste incinerators and mechanical recycling. Red indicates a negative environmental potential. Green indicates a positive environmental potential. White indicates environmental potentials close to zero. Grey indicates that chemical recycling does not exist or has been omitted. PET can be used to produce ethylene glycol and two types of monomers: terephthalic acid (denoted PET TPA) and dimethyl terephthalate denoted PET DMT).

negative potential	positive potential	Marine eutrophication [kg N-Equiv.]					
		refinery feedstock	gaseous fuels	gasoline	diesel	monomers	chemical upcycling
(1) energy recovery in waste incinerator	PET TPA			-2.70E-04	-3.04E-04	6.39E-04	9.48E-04
	PET DMT					7.33E-04	
	LDPE	-8.77E-04	-5.67E-04	-5.52E-04	-6.17E-04	-5.55E-04	
	HDPE	-8.96E-04	-5.78E-04	-5.64E-04	-6.29E-04	-5.67E-04	
	PP	-8.36E-04	-5.31E-04	-5.17E-04	-5.82E-04	-5.32E-04	
	PS	-7.93E-04	-5.62E-04	-5.48E-04	-6.08E-04	5.98E-04	
(2) energy recovery in cement kiln (natural gas)	PET TPA			-1.43E-04	-1.77E-04	7.65E-04	1.07E-03
	PET DMT					8.59E-04	
	LDPE	-5.55E-04	-2.45E-04	-2.30E-04	-2.95E-04	-2.33E-04	
	HDPE	-5.64E-04	-2.47E-04	-2.32E-04	-2.98E-04	-2.35E-04	
	PP	-5.24E-04	-2.19E-04	-2.05E-04	-2.70E-04	-2.20E-04	
	PS	-4.88E-04	-2.57E-04	-2.44E-04	-3.03E-04	9.03E-04	
(3) mechanical recycling	PET TPA			-1.13E-03	-1.17E-03	-2.25E-04	8.41E-05
	PET DMT					-1.31E-04	
	LDPE	9.04E-05	4.01E-04	4.16E-04	3.51E-04	4.12E-04	
	HDPE	2.41E-05	3.42E-04	3.56E-04	2.91E-04	3.53E-04	
	PP	6.20E-06	3.11E-04	3.25E-04	2.61E-04	3.11E-04	
	PS	-7.23E-04	-4.92E-04	-4.79E-04	-5.39E-04	6.67E-04	

Figure A.11: Environmental potential in terms of marine eutrophication for chemical recycling compared to energy recovery in cement kilns that uses lignite as fuel as well as energy recovery in municipal waste incinerators and mechanical recycling. Red indicates a negative environmental potential. Green indicates a positive environmental potential. White indicates environmental potentials close to zero. Grey indicates that chemical recycling does not exist or has been omitted. PET can be used to produce ethylene glycol and two types of monomers: terephthalic acid (denoted PET TPA) and dimethyl terephthalate (denoted PET DMT).

negative potential	positive potential	Freshwater eutrophication [kg P eq]					
		refinery feedstock	gaseous fuels	gasoline	diesel	monomers	chemical upcycling
(1) energy recovery in waste incinerator	PET TPA			-3.66E-04	-3.75E-04	1.22E-04	1.27E-04
	PET DMT					3.80E-05	
	LDPE	-7.66E-04	-7.93E-04	-7.18E-04	-7.36E-04	-7.94E-04	
	HDPE	-7.84E-04	-8.11E-04	-7.35E-04	-7.53E-04	-8.12E-04	
	PP	-7.45E-04	-7.73E-04	-6.98E-04	-7.15E-04	-7.71E-04	
	PS	-7.29E-04	-7.59E-04	-6.90E-04	-7.06E-04	1.96E-04	
(2) energy recovery in cement kiln (natural gas)	PET TPA			-1.98E-03	-1.98E-03	-1.49E-03	-1.48E-03
	PET DMT					-1.57E-03	
	LDPE	-3.61E-03	-3.63E-03	-3.56E-03	-3.58E-03	-3.63E-03	
	HDPE	-3.68E-03	-3.71E-03	-3.64E-03	-3.65E-03	-3.71E-03	
	PP	-3.52E-03	-3.55E-03	-3.48E-03	-3.49E-03	-3.55E-03	
	PS	-3.46E-03	-3.49E-03	-3.42E-03	-3.44E-03	-2.54E-03	
(3) mechanical recycling	PET TPA			-4.12E-04	-4.21E-04	7.67E-05	8.11E-05
	PET DMT					-7.67E-06	
	LDPE	3.09E-04	2.82E-04	3.57E-04	3.40E-04	2.81E-04	
	HDPE	1.95E-04	1.67E-04	2.43E-04	2.26E-04	1.66E-04	
	PP	1.79E-04	1.52E-04	2.27E-04	2.09E-04	1.53E-04	
	PS	-5.50E-04	-5.80E-04	-5.11E-04	-5.27E-04	3.75E-04	

Figure A.12: Environmental potential in terms of freshwater eutrophication for chemical recycling compared to energy recovery in cement kilns that uses lignite as fuel as well as energy recovery in municipal waste incinerators and mechanical recycling. Red indicates a negative environmental potential. Green indicates a positive environmental potential. White indicates environmental potentials close to zero. Grey indicates that chemical recycling does not exist or has been omitted. PET can be used to produce ethylene glycol and two types of monomers: terephthalic acid (denoted PET TPA) and dimethyl terephthalate (denoted PET DMT).

Appendix B

Supplementary information Chapter 4

B.1 Table of life cycle assessment datasets

Table B.1: Life cycle inventory data collected from literature.

Process	Data source
CO_2 source	Ecoinvent and Farla et al., 1999
Polyol production energy and utility demands	Von der Assen et al., 2013
DMC catalyst production	Von der Assen et al., 2013
HNBR production	Own calculations based on Happ et al.
Polyol production and chain extension	Pilot plant datasets

Table B.2: Datasets used from the GABI database (Sphera, 2019).

Product	**Name of dataset**	**Country**
Maleic anhydride	DE: Maleic anhydride ts	DE
Propylene glycol	DE: Propylene glycol ts	DE
Hexamethylene diisocyanate (HDI)	DE: Hexamethylene diisocyanate (HMDI) by-product HCl ts	DE
Process water	EU-27: Process water ts	EU-27
Electricity	DE: Electricity grid mix ts	DE
Compressed air	EU-27: Compressed air ts	EU-27
Nitrogen	EU-27: Nitrogen ts	EU-27
Process steam	EU-27: Process steam from natural gas 90 % ts	EU-27
Waste water treatment	EU-27: Waste water treatment	EU-27
NBR	DE: Nitrile butadiene rubber (NBR, 33 % acrylonitrile) ts	DE
EPDM	DE: Ethylene Propylene Diene Rubber (EPDM) ts	DE
CR	DE: Chloroprene ts	DE

B.2 Elementary composition of CO_2-based and conventional rubbers

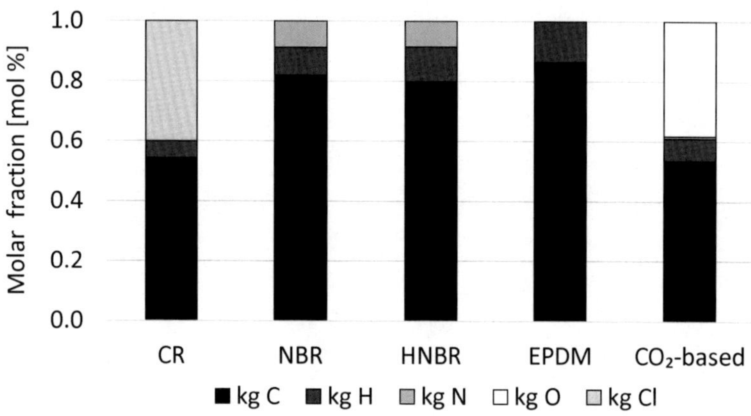

Figure B.1: Composition of CO_2-based and 4 conventional rubbers: hydrated nitrile butadiene rubber (HNBR), nitrile butadiene rubber (NBR), ethylene propylene diene rubber (EPDM) and polychloroprene (CR).

B.3 Midpoint results for conventional and CO_2-based rubbers

B.3.1 Freshwater eutrophication

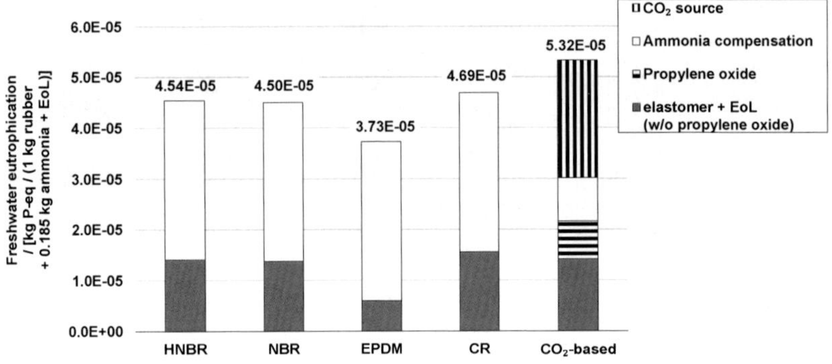

Figure B.2: Freshwater eutrophication in kg phosphorous (P) equivalents of product systems for conventional and CO_2-based rubbers. The CO_2 source includes environmental impacts of the ammonia plant including CO_2 capture, CO_2 compression and CO_2 transport.

B.3.2 Stratospheric ozone depletion

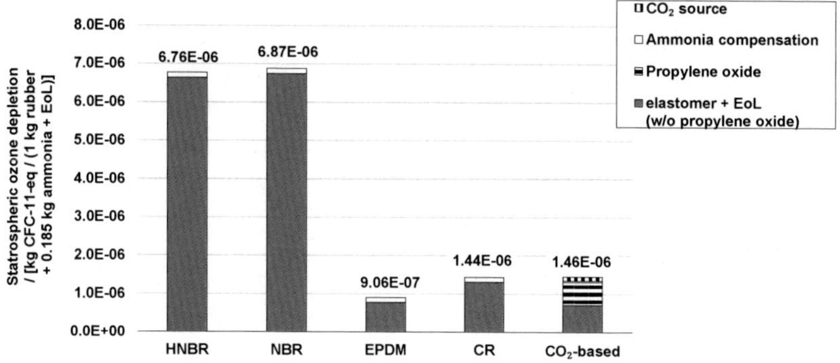

Figure B.3: Stratospheric ozone depletion in kg CFC-11-equivalents of product systems for conventional and CO_2-based rubbers. The CO_2 source includes environmental impacts of the ammonia plant including CO_2 capture, CO_2 compression and CO_2 transport.

B.3.3 Particulate matter formation

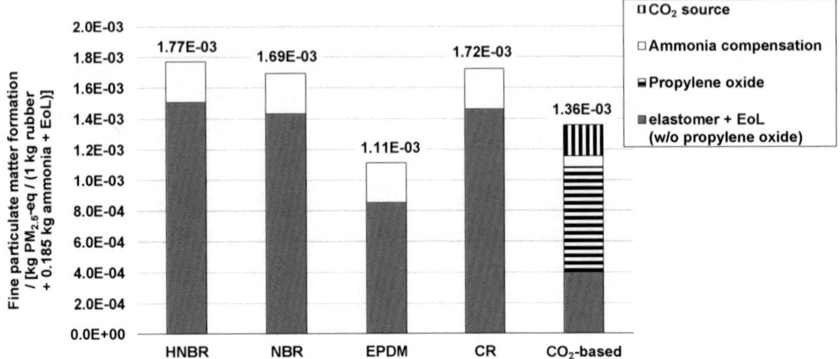

Figure B.4: Fine particulate matter formation in kg PM2.5-equivalents of product systems for conventional and CO_2-based rubbers. The CO_2 source includes environmental impacts of the ammonia plant including CO_2 capture, CO_2 compression and CO_2 transport.

B.3.4 Photochemical ozone formation

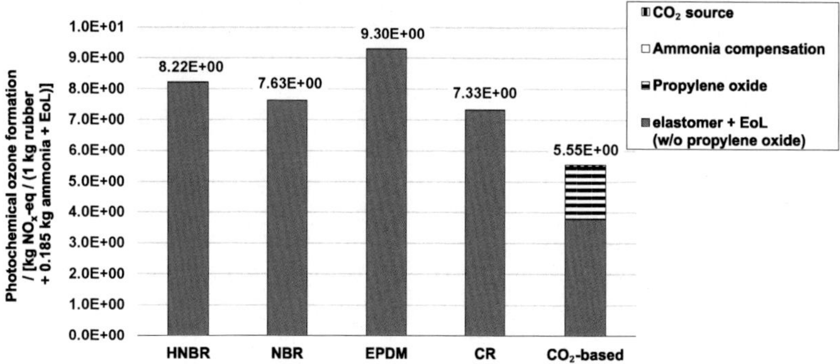

Figure B.5: Photochemical ozone formation in kg NOx-equivalents of product systems for conventional and CO_2-based rubbers. The CO_2 source includes environmental impacts of the ammonia plant including CO_2 capture, CO_2 compression and CO_2 transport.

B.3.5 Terrestrial acidification

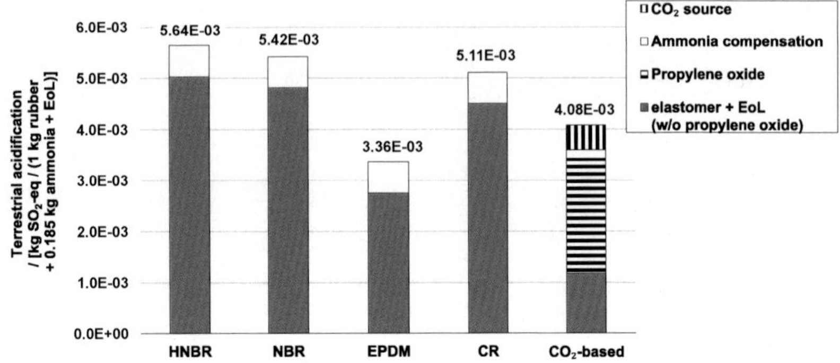

Figure B.6: Terrestrial acidification in kg SO_2-equivalents of product systems for conventional and CO_2-based rubbers. The CO_2 source includes environmental impacts of the ammonia plant including CO_2 capture plus CO_2 compression and CO_2 transport.

B.3.6 Marine eutrophication

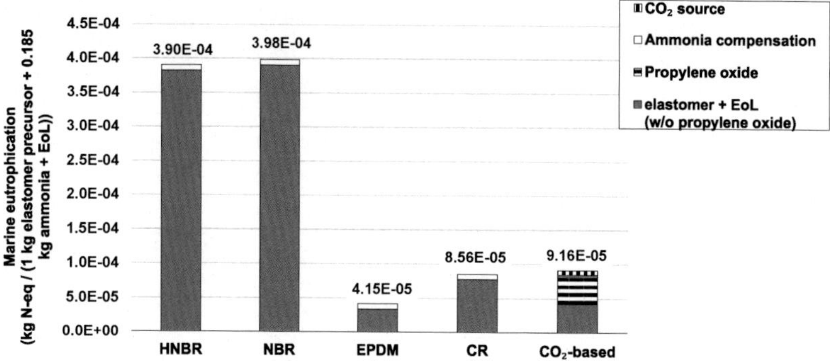

Figure B.7: Marine eutrophication in kg nitrogen (N) equivalents of product systems for conventional and CO_2-based rubbers. The CO_2 source includes environmental impacts of the ammonia plant including CO_2 capture plus CO_2 compression and CO_2 transport.

B.4 Product-specific global warming impacts

The functional unit of this life cycle assessment study includes three services: (i) the production of rubbers, (ii) the production of ammonia as the co-product of the CO_2-source and (iii) the disposal of rubbers, e.g. their incineration. However, in some cases a product-specific environmental impacts can be required (von der Assen et al., 2013, 2014). To derive such product-specific impacts, all environmental impacts have to be allocated to the rubber or the ammonia production (ILCD, 2010a). This allocation introduces ambiguity to the results.

To illustrate the effect of allocation procedure, two scenarios for CO_2-based rubbers are defined: (i) a worst-case and (ii) a best-case allocation scenario. The worst-case allocation scenario assumes that all environmental impacts of CO_2 compression and transport are allocated solely to the CO_2-based rubber. In the best-case scenario, a credit is given for the amount of produced ammonia. The credit is based on the environmental impacts of a conventional ammonia plant without utilizing CO_2 for rubber production. Incineration impacts are not affected in both scenarios.

The global warming impacts of CO_2-based rubbers range between 4.83 kg CO_2-eq for the worst-case and 4.57 kg CO_2-eq for the best-case allocation (Figure B.8, right side). Conventional rubbers range from 5.67 kg CO_2-eq for CR to 7.07 kg CO_2-eq for HNBR. CO_2-based rubbers reduce global warming impacts between 15 % (worst-case, CR) and 35 % (best-case, HNBR). Thus, CO_2-based rubbers reduce global warming impacts even in the worst-case allocation scenario.

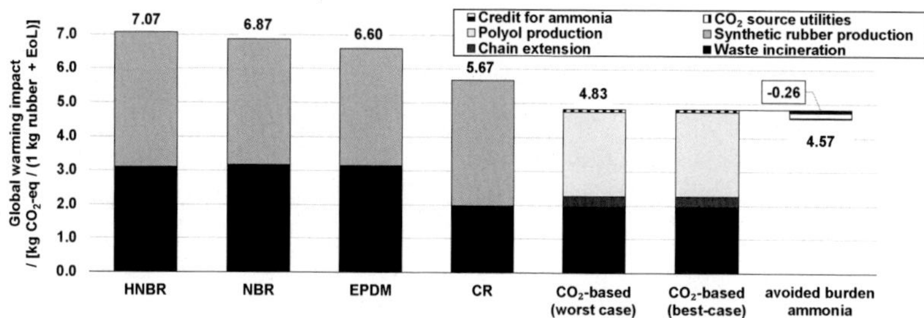

Figure B.8: Product specific global warming impacts are shown in kg CO_2-equivalents per 1 kg of rubber and the end-of-life treatment of 1 kg rubber. For CO_2-based rubbers, environmental impacts were obtained by (i) a worst-case scenario (100 % of environmental impacts allocated to the CO_2-based rubbers) and (ii) a best-case scenario (credit for ammonia). The CO_2 source utilities include CO_2 compression and CO_2 transport.

B.5 Influence of end-of-life allocation

This life cycle assessment study uses the so-called cut-off approach (Ekvall and Tillman, 1997) to deal with the multifunctionality of waste treatment. The cut-off approach assumes that no environmental impacts are allocated to products of waste treatment. Thus, this assumption allocates all environmental impacts of waste treatment to the product under study and can be seen as a worst-case allocation from the perspective of rubber products. However, waste incineration has been shown to produce electricity and heat, especially if high calorific wastes like polymers are incinerated (Eriksson and Finnveden, 2009). In life cycle assessment literature, the by-products of waste treatment are frequently credited by their conventional production technology. This approach is frequently referred to as the avoided-burden approach (ILCD, 2010a) and can be seen as a best-case scenario for rubber products.

To highlight the effect of allocation procedure for waste treatment products, the global warming impact and fossil resource depletion of CO_2-based and conventional rubbers are calculated for the avoided burden approach. For this purpose, an overall energy efficiency of 41 % for the energy recovery (Eriksson and Finnveden, 2009) from polymer incineration is assumed. The share of produced heat is assumed 72 % based on data for average municipal solid waste incinerators in Europe (Eriksson and Finnveden, 2009). The given credit is based on the average German electricity grid and the supply of thermal energy based on natural gas (Sphera, 2019). To calculate the amount of produced electricity and heat, the lower heating values of CO_2-based and conventional rubbers are required. Unfortunately, no lower heating values could be obtained from scientific literature. To fill the data gap, the higher heating values of CO_2-based and conventional rubbers are estimated based on the Vondracek equation and a correction of the obtained higher heating values with the standard enthalpy of evaporation of formed water (44 kJ/mol (Linstrom, 2018)). The Vondracek equation has been shown to have the highest accuracy when estimating higher heating values of waste components (Buckley, 1991). The calculated lower heating values range between 21.7 MJ/kg for CO_2-based rubbers to 42.78 MJ/kg for EPDM.

The results for the avoided burden approach with regard to the global warming impact can be found in Figure B.9. The same recipe for cross-linkable polyether carbonate polyols as in the main paper is used: approx. 20 % CO2, approx. 2 % double bond moieties and a molar mass of polyols of approx. 4200 g/mol. Global warming impacts of CO_2-based rubbers equal 4.08 kg CO_2-eq, if a credit is given for electricity and heat produced from waste incineration. For conventional rubbers, global warming impacts range between 5.06 kg CO_2-eq for CR and 5.90 kg CO_2-eq for HNBR. Credits for electricity and heat production range from 1.67 kg CO_2-eq for EPDM to 0.85 kg CO_2-eq for CO_2-based rubbers. CO_2-based rubbers have smaller lower heating values due to increased oxygen content (cf. Figur B.1) and, thus, CO_2-based rubbers have lower credits for electricity and heat production. However, CO_2-based rubbers reduce global warming impacts between 19 % for CR and 31 % for HNBR. For fossil resource depletion, reductions range between 15 % for CR and 25 % for HNBR.

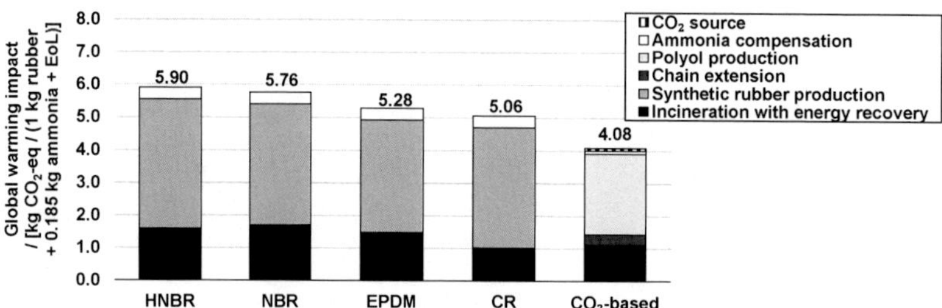

Figure B.9: Global warming impacts in kg CO_2-equivalents of product systems for conventional and CO_2-based rubbers. The CO_2 source utilities include CO_2 compression and CO_2 transport. Here, a credit is given for produced electricity and heat from waste incineration and subtracted from waste incineration emissions.

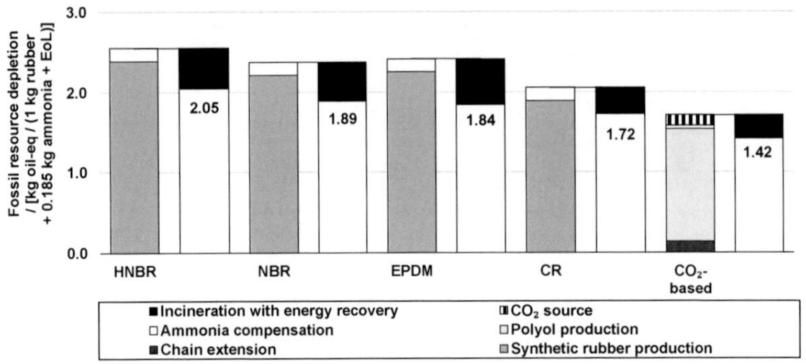

Figure B.10: Fossil resource depletion in kg oil-equivalents of product systems for conventional and CO_2-based rubbers. The CO_2 source utilities include CO_2 compression and CO_2 transport. Here, a credit is given for produced electricity and heat from waste incineration (incineration with energy recovery). For conventional rubbers, the final fossil resource depletion is calculated by subtracting the credit for energy recovery from the fossil resource depletion of rubber production and the ammonia compensation. In case of CO_2-based rubbers, the credit is subtracted from the fossil resource depletion of chain extension, polyol production, ammonia compensation and the CO_2 source.

Appendix C

Documentation of data sources and additional results

This appendix summarizes the data sources to build the bottom-up model explained in Chapter 5.

C.1 Full documentation of data sources

Table C.1: Production technologies and literature sources are shown for each chemical product.

name of flow	technologies	source
acetic acid	carbonylation of methanol	IHS Markit (2018)
acrylic acid, ester grade	from propylene by ammoxidation	IHS Markit (2018)
acrylonitrile	propylene ammoxidation	IHS Markit (2018)
adipic acid	benzene oxidation via cyclohexanol	IHS Markit (2018)
allyl chloride	propylene chlorination	IHS Markit (2018)
aniline	reduction of nitrobenzene	IHS Markit (2018)
caprolactam	production from toluene	IHS Markit (2018)
carbon monoxide	partial condensation of synthesis gas	IHS Markit (2018)
chlorine	electrolysis of hydrochloric acid	IHS Markit (2018)
chlorine	electrolysis of NaCl in membrane cell	IHS Markit (2018)
chlorine	electrolysis of NaCl in diaphragm cell	IHS Markit (2018)
chlorine	electrolysis of NaCl in mercury cell	IHS Markit (2018)
chlorine	electrolysis via oxygen-depolarized cathodes	IHS Markit (2018)
polyamide 6	continous PA 6 production	IHS Markit (2018)
cumene	alkylation of benzene with propylene	IHS Markit (2018)
cyclohexane	hydrogenation of benzene	IHS Markit (2018)
dimethyl terephthalate	esterification of terephthalic acid with methanol	IHS Markit (2018)
dinitrotoluene	nitration of toluene	IHS Markit (2018)
epichlorohydrin	chlorohydrination of allyl chloride	IHS Markit (2018)
ethanol	catalytic hydration of ethylene (not included in the final assessment)	IHS Markit (2018)

continued on next page

Table C.1 – *continued from previous page*

name of flow	technology	sources
ethylbenzene	alkylation of benzene with ethylene (zeolite catalyst)	IHS Markit (2018)
ethylene	UOP/HYDRO methanol to olefins	IHS Markit (2018)
ethylene	ethane steam cracking (not included in the final assessment)	IHS Markit (2018)
ethylene	catalytic dehydration of ethanol (adiabatic fixed-bed)	IHS Markit (2018)
ethylene	catalytic dehydration of ethanol (fluidized-bed)	IHS Markit (2018)
ethylene glycol	thermal hydration of ethylene oxide	IHS Markit (2018)
formaldehyde	oxidation of methanol (ferric-molybdate catalyst)	IHS Markit (2018)
formaldehyde	oxidation of methanol (silver cat.)	IHS Markit (2018)
glycerin	oxidation of allyl chloride via epichlorohydrin	IHS Markit (2018)
hexamethylene-diamine	from acrylonitrile via diponitrile	IHS Markit (2018)
hexamethylene-diamine	from acrylonitrile via diponitrile (electrohydrodimerization)	IHS Markit (2018)
methanol	from synthesis gas	IHS Markit (2018)
ethylene	DMTO methanol to olefins (C2/C3 = 1.5)	IHS Markit (2018)
propylene	Lurgi methanol to propylene	IHS Markit (2018)
methyl acrylate	esterification of acrylic acid	IHS Markit (2018)
methylene diphenyl diisocyanate	phosgenation of benzene	IHS Markit (2018)
nitric acid (60%)	from ammonia (dual pressure)	IHS Markit (2018)
nitric acid (60%)	from ammonia (mono pressure)	IHS Markit (2018)
nitrobenzene	nitration of benzene (adiabatic)	IHS Markit (2018)
nitrobenzene	nitration of benzene (conventional)	IHS Markit (2018)
nitrogen	air separation by pressure-swing adsorption	IHS Markit (2018)
polyamide 66	continous PA 66 production	IHS Markit (2018)
oxygen	cryogenic air separation	IHS Markit (2018)
phenol	oxidation of cumene	IHS Markit (2018)
polyacrylonitrile fiber	melt extrusion	IHS Markit (2018)
polybutadiene	solution polymerization	IHS Markit (2018)
PET pellets (fiber-grade)	polycondensation from dimethyl terephthalate and ethylene glycol	IHS Markit (2018)
PET pellets (fiber-grade)	polycondensation from terephthalic acid and ethylene glycol	IHS Markit (2018)
PET pellets (bottle-grade)	upgrading of PET fiber-grade to bottle-grade	IHS Markit (2018)
polyethylene, HD	gas-phase polymerization	IHS Markit (2018)
polyethylene, LD	autoclave polymerization	IHS Markit (2018)
polyethylene, LLD	solution polymerization	IHS Markit (2018)
polyol (polyester-based)	from adipic acid and diethylene glycol	IHS Markit (2018)
polyol (polyether-based)	from propylene/ethylene oxide and glycerol	IHS Markit (2018)
polypropylene	gas-phase polymerization	IHS Markit (2018)
polystyrene, GP	bulk polymerization	IHS Markit (2018)
polystyrene, HI	bulk polymerization (incl. Polybutadiene)	IHS Markit (2018)
polyvinyl chloride	suspension polymerization	IHS Markit (2018)
propylene	dimerization of ethylene	IHS Markit (2018)
propylene glycol	propylene oxide oxidation	IHS Markit (2018)
propylene oxide	chlorohydrine process	IHS Markit (2018)
propylene	ethylene disproportionation	IHS Markit (2018)

continued on next page

Table C.1 – *continued from previous page*

name of flow	technology	sources
steam	natural gas boiler	IHS Markit (2018)
styrene	alkylation of benzene with ethylene (liquid-phase)	IHS Markit (2018)
styrene	alkylation of benzene with ethylene (gas-phase)	IHS Markit (2018)
synthesis gas (2:1)	natural gas steam reforming	IHS Markit (2018)
terephthalic acid	oxidation of p-xylene	IHS Markit (2018)
toluene diisocyanate	phosgenation of toluene	IHS Markit (2018)
vinyl chloride	ethylene chlorination and ethylene dichloride pyrolysis	IHS Markit (2018)
ammonia	integrated process of steam methane reforming and Haber-Bosch process	IHS Markit (2018)
ethylene oxide	ethylene oxidation with oxygen	IHS Markit (2018)
hydrogen	steam methane reforming and water-gas-shift	IHS Markit (2018)
carbon dioxide	compression to 100bar	Farla et al., 1995
benzene	solvent extraction from reformate	IHS Markit (2018)
toluene	solvent extraction from reformate	IHS Markit (2018)
xylenes, mixed	solvent extraction from reformate	IHS Markit (2018)
benzene	solvent extraction from pyrolysis gasoline	IHS Markit (2018)
toluene	solvent extraction from pyrolysis gasoline	IHS Markit (2018)
xylenes, mixed	solvent extraction from pyrolysis gasoline	IHS Markit (2018)
benzene	separation of xylenes by adsorption	IHS Markit (2018)
toluene	separation of xylenes by adsorption	IHS Markit (2018)
p-xylene	separation of xylenes by adsorption	IHS Markit (2018)
o-xylene	separation of xylenes by adsorption	IHS Markit (2018)
benzene	separation of xylenes by crystallization	IHS Markit (2018)
toluene	separation of xylenes by crystallization	IHS Markit (2018)
p-xylene	separation of xylenes by crystallization	IHS Markit (2018)
o-xylene	separation of xylenes by crystallization	IHS Markit (2018)
pyrolysis gasoline	steam cracking of naphtha	IHS Markit (2018)
ethylene	steam cracking of naphtha	IHS Markit (2018)
propylene	steam cracking of naphtha	IHS Markit (2018)
polyurethane, flexible	continous production of flexible polyurethane (ecoinvent)	Ecoinvent (2020)
polyurethane, rigid	continous production of flexible polyurethane foam (ecoinvent)	Ecoinvent (2020)
thermal energy	thermal energy recovery from hydrocarbon fuels	own calculation: based on net calorific values
thermal energy	thermal energy from natural gas	own calculation: based on net calorific values
oleum (33.3%)	33.3% oleum from sulfur trioxide and sulfuric acid	own calculation: 0.333 kg sulfur trioxide + 0.667 kg sulfuric acid
synthesis gas (2:1)	mixing of hydrogen and CO (2:1)	own calculation: 0.875 kg CO + 0.125 kg H2
hydrogen	water-gas shift and amine separation of CO2	IHS Markit (2018)
sodium chloride	sodium chloride production, powder (ecoinvent)	Ecoinvent (2020)
electricity	electricity sink without avoided burden	assumption: no avoided burden only sink
cyclohexane	waste treatment of cyclohexane	own calculation: Appendix A
diethylene glycol	waste treatment of diethylene glycol (crude)	own calculation: Appendix A
dichloropropylenes	waste treatment of dichloropropylenes	own calculation: Appendix A
residual waste methanol to BTX	waste treatment of residual waste methanol to BTX	own calculation: Appendix A

continued on next page

Table C.1 – *continued from previous page*

name of flow	technology	sources
residual waste methanol to BTX	waste treatment of residual waste methanol to BTX (with CO2 capture)	own calculation: Appendix A
diethylene glycol	waste treatment of diethylene glycol	own calculation: Appendix A
caustic soda (50%)	waste treatment of caustic soda (50%)	own calculation: Appendix A
hydrogen cyanide	waste treatment of hydrogen cyanide	own calculation: Appendix A
hydrochloric acid	waste treatment of hydrochloric acid	own calculation: Appendix A
chlorine	waste treatment of chlorine	own calculation: Appendix A
benzene	methanol to aromatics	IHS Markit (2018)
toluene	methanol to aromatics	IHS Markit (2018)
xylenes, mixed	methanol to aromatics	IHS Markit (2018)
p-xylene	methanol to aromatics	IHS Markit (2018)
ethylene	DMTO methanol to olefins (C2/C3 = 2.5)	IHS Markit (2018)
natural gas (raw material)	resistance heater	own calculation: Appendix C.2
ammonia	Haber-Bosch process	Cesaro et al. (2021)
hydrogen	low-temperature electrolysis	Agora (2018)
steam	steam sink without avoided burden	assumption: no avoided burden only sink
steam	electric boiler	own calculation: Appendix C.2
methanol	thermal hydrogenation	Pérez-Fortes and Tzimas (2016)
methanol	thermal hydrogenation with CO2 recycle	Pérez-Fortes and Tzimas (2016)
benzene	9 CO2 + 27 H2 → C6H6 + 18 H2O + 3 CH4 (early-stage)	own calculation: Chapter 5.4
toluene	9 CO2 + 26 H2 → C7H8 + 18 H2O + 2 CH4 (early-stage)	own calculation: Chapter 5.4
xylenes, mixed	9 CO2 + 25 H2 → mixed C8H10 + 18 H2O + CH4 (early-stage)	own calculation: Chapter 5.4
p-xylene	9 CO2 + 25 H2 → para C8H10 + 18 H2O + CH4 (early-stage)	own calculation: Chapter 5.4
carbon monoxide	CO2 + H2 → CO + H2O (early-stage)	own calculation: Chapter 5.4
ethylene oxide	C2H4 + CO2 → C2H4O + CO (early-stage)	own calculation: Chapter 5.4
ethylene	2 CO2 + 2 CH4 → C2H4 + 2 CO + 2 H2O (early-stage)	own calculation: Chapter 5.4
ethylene	2 CO2 + 6 H2 → C2H4 + 4 H2O (early-stage)	own calculation: Chapter 5.4
propylene	3 CO2 + 9 H2 → C3H6 + 6 H2O (early-stage)	own calculation: Chapter 5.4
styrene	C8H10 + CO2 → C8H8 + CO + H2O (early-stage)	own calculation: Chapter 5.4
ethanol	hydrolysis and fermentation (wood pellets)	Humbird et al. (2011)
ethanol	hydrolysis and fermentation (wood chips)	Humbird et al. (2011)
ethanol	hydrolysis and fermentation (miscanthus)	Humbird et al. (2011)
synthesis gas (2:1)	gasification of biomass (wood chips, WGS)	Arvidsson et al. (2014); Arvidsson, M. et al. (2012); Hannula and Kurkela (2010); Isaksson et al. (2012); Spath et al. (2005)
synthesis gas (2:1)	gasification of biomass (wood chips, H2)	Arvidsson et al. (2014); Arvidsson, M. et al. (2012); Hannula and Kurkela (2010); Isaksson et al. (2012); Spath et al. (2005)

continued on next page

Table C.1 – *continued from previous page*

name of flow	technology	sources
synthesis gas (2:1)	gasification of biomass (bark chips, WGS)	Arvidsson et al. (2014); Arvidsson, M. et al. (2012); Hannula and Kurkela (2010); Isaksson et al. (2012); Spath et al. (2005)
synthesis gas (2:1)	gasification of biomass (bark chips, H2)	Arvidsson et al. (2014); Arvidsson, M. et al. (2012); Hannula and Kurkela (2010); Isaksson et al. (2012); Spath et al. (2005)
synthesis gas (2:1)	gasification of biomass (miscanthus, WGS)	Arvidsson et al. (2014); Arvidsson, M. et al. (2012); Hannula and Kurkela (2010); Isaksson et al. (2012); Spath et al. (2005)
synthesis gas (2:1)	gasification of biomass (miscanthus, H2)	Arvidsson et al. (2014); Arvidsson, M. et al. (2012); Hannula and Kurkela (2010); Isaksson et al. (2012); Spath et al. (2005)
synthesis gas (2:1)	gasification of biomass (wood pellets, WGS)	Arvidsson et al. (2014); Arvidsson, M. et al. (2012); Hannula and Kurkela (2010); Isaksson et al. (2012); Spath et al. (2005)
synthesis gas (2:1)	gasification of biomass (wood pellets, H2)	Arvidsson et al. (2014); Arvidsson, M. et al. (2012); Hannula and Kurkela (2010); Isaksson et al. (2012); Spath et al. (2005)
synthesis gas (2:1)	gasification of biomass (wood chips, WGS)	Arvidsson et al. (2014); Arvidsson, M. et al. (2012); Hannula and Kurkela (2010); Isaksson et al. (2012); Spath et al. (2005)
synthesis gas (2:1)	gasification of biomass (wood chips, H2)	Arvidsson et al. (2014); Arvidsson, M. et al. (2012); Hannula and Kurkela (2010); Isaksson et al. (2012); Spath et al. (2005)
synthesis gas (2:1)	gasification of biomass (bark chips, WGS)	Arvidsson et al. (2014); Arvidsson, M. et al. (2012); Hannula and Kurkela (2010); Isaksson et al. (2012); Spath et al. (2005)
synthesis gas (2:1)	gasification of biomass (bark chips, H2)	Arvidsson et al. (2014); Arvidsson, M. et al. (2012); Hannula and Kurkela (2010); Isaksson et al. (2012); Spath et al. (2005)
synthesis gas (2:1)	gasification of biomass (miscanthus, WGS)	Arvidsson et al. (2014); Arvidsson, M. et al. (2012); Hannula and Kurkela (2010); Isaksson et al. (2012); Spath et al. (2005)

continued on next page

Table C.1 – *continued from previous page*

name of flow	technology	sources
synthesis gas (2:1)	gasification of biomass (miscanthus, H2)	Arvidsson et al. (2014); Arvidsson, M. et al. (2012); Hannula and Kurkela (2010); Isaksson et al. (2012); Spath et al. (2005)
synthesis gas (2:1)	gasification of biomass (wood pellets, WGS)	Arvidsson et al. (2014); Arvidsson, M. et al. (2012); Hannula and Kurkela (2010); Isaksson et al. (2012); Spath et al. (2005)
synthesis gas (2:1)	gasification of biomass (wood pellets, H2)	Arvidsson et al. (2014); Arvidsson, M. et al. (2012); Hannula and Kurkela (2010); Isaksson et al. (2012); Spath et al. (2005)
steam	from wood chips	Pérez-Uresti et al. (2019)
steam	from bark chips	Pérez-Uresti et al. (2019)
steam	from miscanthus	Pérez-Uresti et al. (2019)
steam	from wood pellets	Pérez-Uresti et al. (2019)
PP packaging waste	monomer recovery from plastic waste (early-stage)	own calculation: Chapter 5.4
LDPE packaging waste	monomer recovery from plastic waste (early-stage)	own calculation: Chapter 5.4
LLDPE packaging waste	monomer recovery from plastic waste (early-stage)	own calculation: Chapter 5.4
HDPE packaging waste	monomer recovery from plastic waste (early-stage)	own calculation: Chapter 5.4
PET packaging waste	monomer recovery from plastic waste (early-stage)	own calculation: Chapter 5.4
GPPS packaging waste	monomer recovery from plastic waste (early-stage)	own calculation: Chapter 5.4
HIPS packaging waste	monomer recovery from plastic waste (early-stage)	own calculation: Chapter 5.4
PP waste	monomer recovery from plastic waste (early-stage)	own calculation: Chapter 5.4
LDPE waste	monomer recovery from plastic waste (early-stage)	own calculation: Chapter 5.4
LLDPE waste	monomer recovery from plastic waste (early-stage)	own calculation: Chapter 5.4
HDPE waste	monomer recovery from plastic waste (early-stage)	own calculation: Chapter 5.4
PET waste	monomer recovery from plastic waste (early-stage)	own calculation: Chapter 5.4
PP&A - PET waste	monomer recovery from plastic waste (early-stage)	own calculation: Chapter 5.4
GPPS waste	monomer recovery from plastic waste (early-stage)	own calculation: Chapter 5.4
HIPS waste	monomer recovery from plastic waste (early-stage)	own calculation: Chapter 5.4
PVC waste	monomer recovery from plastic waste (early-stage)	own calculation: Chapter 5.4
PUR flex waste	monomer recovery from plastic waste (early-stage)	own calculation: Chapter 5.4
PUR rigid waste	monomer recovery from plastic waste (early-stage)	own calculation: Chapter 5.4
PP&A - polyamide 6 waste	monomer recovery from plastic waste (early-stage)	own calculation: Chapter 5.4
PP&A - polyamide 66 waste	monomer recovery from plastic waste (early-stage)	own calculation: Chapter 5.4
PP&A - acrylic waste	monomer recovery from plastic waste (early-stage)	own calculation: Chapter 5.4
LLDPE (DSD spec 310)	monomer recovery from plastic waste (early-stage)	own calculation: Chapter 5.4

continued on next page

Table C.1 – *continued from previous page*

name of flow	technology	sources
LDPE (DSD spec 310)	monomer recovery from plastic waste (early-stage)	own calculation: Chapter 5.4
HDPE (DSD spec 329)	monomer recovery from plastic waste (early-stage)	own calculation: Chapter 5.4
PP (DSD spec 324)	monomer recovery from plastic waste (early-stage)	own calculation: Chapter 5.4
GPPS (DSD spec 331)	monomer recovery from plastic waste (early-stage)	own calculation: Chapter 5.4
HIPS (DSD spec 331)	monomer recovery from plastic waste (early-stage)	own calculation: Chapter 5.4
PET (DSD spec 325)	monomer recovery from plastic waste (early-stage)	own calculation: Chapter 5.4
rest LLDPE (DSD spec 310)	monomer recovery from plastic waste (early-stage)	own calculation: Chapter 5.4
rest LDPE (DSD spec 310)	monomer recovery from plastic waste (early-stage)	own calculation: Chapter 5.4
rest HDPE (DSD spec 329)	monomer recovery from plastic waste (early-stage)	own calculation: Chapter 5.4
rest PP (DSD spec 324)	monomer recovery from plastic waste (early-stage)	own calculation: Chapter 5.4
rest GPPS (DSD spec 331)	monomer recovery from plastic waste (early-stage)	own calculation: Chapter 5.4
rest HIPS (DSD spec 331)	monomer recovery from plastic waste (early-stage)	own calculation: Chapter 5.4
rest PET (DSD spec 325)	monomer recovery from plastic waste (early-stage)	own calculation: Chapter 5.4
MR residual of LLDPE (DSD spec 310)	monomer recovery from plastic waste (early-stage)	own calculation: Chapter 5.4
MR residual of LDPE (DSD spec 310)	monomer recovery from plastic waste (early-stage)	own calculation: Chapter 5.4
MR residual of HDPE (DSD spec 329)	monomer recovery from plastic waste (early-stage)	own calculation: Chapter 5.4
MR residual of PP (DSD spec 324)	monomer recovery from plastic waste (early-stage)	own calculation: Chapter 5.4
MR residual of GPPS (DSD spec 331)	monomer recovery from plastic waste (early-stage)	own calculation: Chapter 5.4
MR residual of HIPS (DSD spec 331)	monomer recovery from plastic waste (early-stage)	own calculation: Chapter 5.4
MR residual of PET (DSD spec 325)	monomer recovery from plastic waste (early-stage)	own calculation: Chapter 5.4
PET packaging waste	monomer recovery from plastic waste (early-stage)	own calculation: Chapter 5.4
PET waste	monomer recovery from plastic waste (early-stage)	own calculation: Chapter 5.4
PP&A - PET waste	monomer recovery from plastic waste (early-stage)	own calculation: Chapter 5.4

continued on next page

Table C.1 – *continued from previous page*

name of flow	technology	sources
rest LLDPE (DSD spec 310)	monomer recovery from plastic waste (early-stage)	own calculation: Chapter 5.4
rest PET (DSD spec 325)	monomer recovery from plastic waste (early-stage)	own calculation: Chapter 5.4
MR residual of PET (DSD spec 325)	monomer recovery from plastic waste (early-stage)	own calculation: Chapter 5.4
residue LDPE to energy recovery	energy recovery	own calculation: Appendix A
residue HDPE to energy recovery	energy recovery	own calculation: Appendix A
residue PP to energy recovery	energy recovery	own calculation: Appendix A
residue GPPS/HIPS to energy recovery	energy recovery	own calculation: Appendix A
residue PET to TPA to energy recovery	energy recovery	own calculation: Appendix A
residue PET to DMT to energy recovery	energy recovery	own calculation: Appendix A
residue PUR flex to energy recovery	energy recovery	own calculation: Appendix A
residue PUR rigid to energy recovery	energy recovery	own calculation: Appendix A
residue polyamid 6 to energy recovery	energy recovery	own calculation: Appendix A
residue polyamid 66 to energy recovery	energy recovery	own calculation: Appendix A
residue PAN to energy recovery	energy recovery	own calculation: Appendix A
residue PVC to energy recovery	energy recovery	own calculation: Appendix A
residue LDPE to energy recovery	energy recovery	own calculation: Appendix A
residue HDPE to energy recovery	energy recovery	own calculation: Appendix A
residue PP to energy recovery	energy recovery	own calculation: Appendix A
residue GPPS/HIPS to energy recovery	energy recovery	own calculation: Appendix A
residue PET to tpa to energy recovery	energy recovery	own calculation: Appendix A
residue PET to dmt to energy recovery	energy recovery	own calculation: Appendix A

continued on next page

Table C.1 – *continued from previous page*

name of flow	technology	sources
residue PUR flex to energy recovery	energy recovery	own calculation: Appendix A
residue PUR rigid to energy recovery	energy recovery	own calculation: Appendix A
residue polyamid 6 to energy recovery	energy recovery	own calculation: Appendix A
residue polyamid 66 to energy recovery	energy recovery	own calculation: Appendix A
residue PAN to energy recovery	energy recovery	own calculation: Appendix A
residue PVC to energy recovery	energy recovery	own calculation: Appendix A
polyethylene, LLD	mechanical recycling from plastic packaging waste	own calculation: Appendix A
polyethylene, LD	mechanical recycling from plastic packaging waste	own calculation: Appendix A
polyethylene, HD	mechanical recycling from plastic packaging waste	own calculation: Appendix A
polypropylene	mechanical recycling from plastic packaging waste	own calculation: Appendix A
polystyrene, GP	mechanical recycling from plastic packaging waste	own calculation: Appendix A
polystyrene, HI	mechanical recycling from plastic packaging waste	own calculation: Appendix A
PET pellets (fiber-grade)	mechanical recycling from plastic packaging waste	own calculation: Appendix A
naphtha	pyrolysis of PP packaging waste	IHS Markit (2018)
naphtha	pyrolysis of LDPE packaging waste	IHS Markit (2018)
naphtha	pyrolysis of LLDPE packaging waste	IHS Markit (2018)
naphtha	pyrolysis of HDPE packaging waste	IHS Markit (2018)
naphtha	pyrolysis of PET packaging waste	IHS Markit (2018)
naphtha	pyrolysis of GPPS packaging waste	IHS Markit (2018)
naphtha	pyrolysis of HIPS packaging waste	IHS Markit (2018)
naphtha	pyrolysis of PP waste	IHS Markit (2018)
naphtha	pyrolysis of LDPE waste	IHS Markit (2018)
naphtha	pyrolysis of LLDPE waste	IHS Markit (2018)
naphtha	pyrolysis of HDPE waste	IHS Markit (2018)
naphtha	pyrolysis of PET waste	IHS Markit (2018)
naphtha	pyrolysis of PP&A - PET waste	IHS Markit (2018)
naphtha	pyrolysis of GPPS waste	IHS Markit (2018)
naphtha	pyrolysis of HIPS waste	IHS Markit (2018)
naphtha	pyrolysis of PVC waste	IHS Markit (2018)
naphtha	pyrolysis of PUR flex waste	IHS Markit (2018)
naphtha	pyrolysis of PUR rigid waste	IHS Markit (2018)
naphtha	pyrolysis of PP&A - Polyamide 6 waste	IHS Markit (2018)
naphtha	pyrolysis of PP&A - Polyamide 66 waste	IHS Markit (2018)
naphtha	pyrolysis of PP&A - Acrylic waste	IHS Markit (2018)
naphtha	pyrolysis of LLDPE (DSD Spec 310)	IHS Markit (2018)
naphtha	pyrolysis of LDPE (DSD Spec 310)	IHS Markit (2018)
naphtha	pyrolysis of HDPE (DSD Spec 329)	IHS Markit (2018)
naphtha	pyrolysis of PP (DSD Spec 324)	IHS Markit (2018)
naphtha	pyrolysis of GPPS (DSD Spec 331)	IHS Markit (2018)
naphtha	pyrolysis of HIPS (DSD Spec 331)	IHS Markit (2018)
naphtha	pyrolysis of PET (DSD Spec 325)	IHS Markit (2018)
naphtha	pyrolysis of rest LLDPE (DSD Spec 310)	IHS Markit (2018)
naphtha	pyrolysis of rest LDPE (DSD Spec 310)	IHS Markit (2018)

continued on next page

Table C.1 – *continued from previous page*

name of flow	technology	sources
naphtha	pyrolysis of rest HDPE (DSD Spec 329)	IHS Markit (2018)
naphtha	pyrolysis of rest PP (DSD Spec 324)	IHS Markit (2018)
naphtha	pyrolysis of rest GPPS (DSD Spec 331)	IHS Markit (2018)
naphtha	pyrolysis of rest HIPS (DSD Spec 331)	IHS Markit (2018)
naphtha	pyrolysis of rest PET (DSD Spec 325)	IHS Markit (2018)
naphtha	pyrolysis of MR residual of LLDPE (DSD Spec 310)	IHS Markit (2018)
naphtha	pyrolysis of MR residual of LDPE (DSD Spec 310)	IHS Markit (2018)
naphtha	pyrolysis of MR residual of HDPE (DSD Spec 329)	IHS Markit (2018)
naphtha	pyrolysis of MR residual of PP (DSD Spec 324)	IHS Markit (2018)
naphtha	pyrolysis of MR residual of GPPS (DSD Spec 331)	IHS Markit (2018)
naphtha	pyrolysis of MR residual of HIPS (DSD Spec 331)	IHS Markit (2018)
naphtha	pyrolysis of MR residual of PET (DSD Spec 325)	IHS Markit (2018)
LLDPE packaging waste	sorting of plastic packaging waste	own calculation: Appendix A
LDPE packaging waste	sorting of plastic packaging waste	own calculation: Appendix A
HDPE packaging waste	sorting of plastic packaging waste	own calculation: Appendix A
PP packaging waste	sorting of plastic packaging waste	own calculation: Appendix A
GPPS packaging waste	sorting of plastic packaging waste	own calculation: Appendix A
HIPS packaging waste	sorting of plastic packaging waste	own calculation: Appendix A
PET packaging waste	sorting of plastic packaging waste	own calculation: Appendix A
PP packaging waste	energy recovery	own calculation: Appendix A
LDPE packaging waste	energy recovery	own calculation: Appendix A
LLDPE packaging waste	energy recovery	own calculation: Appendix A
HDPE packaging waste	energy recovery	own calculation: Appendix A
PET packaging waste	energy recovery	own calculation: Appendix A
GPPS packaging waste	energy recovery	own calculation: Appendix A
HIPS packaging waste	energy recovery	own calculation: Appendix A
PP waste	energy recovery	own calculation: Appendix A
LDPE waste	energy recovery	own calculation: Appendix A
LLDPE waste	energy recovery	own calculation: Appendix A
HDPE waste	energy recovery	own calculation: Appendix A
PET waste	energy recovery	own calculation: Appendix A
PP&A - PET waste	energy recovery	own calculation: Appendix A
GPPS waste	energy recovery	own calculation: Appendix A
HIPS waste	energy recovery	own calculation: Appendix A
PVC waste	energy recovery	own calculation: Appendix A
PUR flex waste	energy recovery	own calculation: Appendix A
PUR rigid waste	energy recovery	own calculation: Appendix A
PP&A - polyamide 6 waste	energy recovery	own calculation: Appendix A

continued on next page

Table C.1 – *continued from previous page*

name of flow	technology	sources
PP&A - polyamide 66 waste	energy recovery	own calculation: Appendix A
PP&A - acrylic waste	energy recovery	own calculation: Appendix A
LLDPE (DSD spec 310)	energy recovery	own calculation: Appendix A
LDPE (DSD spec 310)	energy recovery	own calculation: Appendix A
HDPE (DSD spec 329)	energy recovery	own calculation: Appendix A
PP (DSD spec 324)	energy recovery	own calculation: Appendix A
GPPS (DSD spec 331)	energy recovery	own calculation: Appendix A
HIPS (DSD spec 331)	energy recovery	own calculation: Appendix A
PET (DSD spec 325)	energy recovery	own calculation: Appendix A
rest LLDPE (DSD spec 310)	energy recovery	own calculation: Appendix A
rest LDPE (DSD spec 310)	energy recovery	own calculation: Appendix A
rest HDPE (DSD spec 329)	energy recovery	own calculation: Appendix A
rest PP (DSD spec 324)	energy recovery	own calculation: Appendix A
rest GPPS (DSD spec 331)	energy recovery	own calculation: Appendix A
rest HIPS (DSD spec 331)	energy recovery	own calculation: Appendix A
rest PET (DSD spec 325)	energy recovery	own calculation: Appendix A
MR residual of LLDPE (DSD spec 310)	energy recovery	own calculation: Appendix A
MR residual of LDPE (DSD spec 310)	energy recovery	own calculation: Appendix A
MR residual of HDPE (DSD spec 329)	energy recovery	own calculation: Appendix A
MR residual of PP (DSD spec 324)	energy recovery	own calculation: Appendix A
MR residual of GPPS (DSD spec 331)	energy recovery	own calculation: Appendix A
MR residual of HIPS (DSD spec 331)	energy recovery	own calculation: Appendix A
MR residual of PET (DSD spec 325)	energy recovery	own calculation: Appendix A

continued on next page

Table C.1 – *continued from previous page*

name of flow	technology	sources
PP packaging waste	energy recovery with CO2 capture	own calculation: Appendix A + Haaf et al. (2020)
LDPE packaging waste	energy recovery with CO2 capture	own calculation: Appendix A + Haaf et al. (2020)
LLDPE packaging waste	energy recovery with CO2 capture	own calculation: Appendix A + Haaf et al. (2020)
HDPE packaging waste	energy recovery with CO2 capture	own calculation: Appendix A + Haaf et al. (2020)
PET packaging waste	energy recovery with CO2 capture	own calculation: Appendix A + Haaf et al. (2020)
GPPS packaging waste	energy recovery with CO2 capture	own calculation: Appendix A + Haaf et al. (2020)
HIPS packaging waste	energy recovery with CO2 capture	own calculation: Appendix A + Haaf et al. (2020)
PP waste	energy recovery with CO2 capture	own calculation: Appendix A + Haaf et al. (2020)
LDPE waste	energy recovery with CO2 capture	own calculation: Appendix A + Haaf et al. (2020)
LLDPE waste	energy recovery with CO2 capture	own calculation: Appendix A + Haaf et al. (2020)
HDPE waste	energy recovery with CO2 capture	own calculation: Appendix A + Haaf et al. (2020)
PET waste	energy recovery with CO2 capture	own calculation: Appendix A + Haaf et al. (2020)
PP&A - PET waste	energy recovery with CO2 capture	own calculation: Appendix A + Haaf et al. (2020)
GPPS waste	energy recovery with CO2 capture	own calculation: Appendix A + Haaf et al. (2020)
HIPS waste	energy recovery with CO2 capture	own calculation: Appendix A + Haaf et al. (2020)
PVC waste	energy recovery with CO2 capture	own calculation: Appendix A + Haaf et al. (2020)
PUR flex waste	energy recovery with CO2 capture	own calculation: Appendix A + Haaf et al. (2020)
PUR rigid waste	energy recovery with CO2 capture	own calculation: Appendix A + Haaf et al. (2020)
PP&A - polyamide 6 waste	energy recovery with CO2 capture	own calculation: Appendix A + Haaf et al. (2020)
PP&A - polyamide 66 waste	energy recovery with CO2 capture	own calculation: Appendix A + Haaf et al. (2020)
PP&A - acrylic waste	energy recovery with CO2 capture	own calculation: Appendix A + Haaf et al. (2020)
LLDPE (DSD spec 310)	energy recovery with CO2 capture	own calculation: Appendix A + Haaf et al. (2020)
LDPE (DSD spec 310)	energy recovery with CO2 capture	own calculation: Appendix A + Haaf et al. (2020)
HDPE (DSD spec 329)	energy recovery with CO2 capture	own calculation: Appendix A + Haaf et al. (2020)
PP (DSD spec 324)	energy recovery with CO2 capture	own calculation: Appendix A + Haaf et al. (2020)
GPPS (DSD spec 331)	energy recovery with CO2 capture	own calculation: Appendix A + Haaf et al. (2020)

continued on next page

Table C.1 – *continued from previous page*

name of flow	technology	sources
HIPS (DSD spec 331)	energy recovery with CO2 capture	own calculation: Appendix A + Haaf et al. (2020)
PET (DSD spec 325)	energy recovery with CO2 capture	own calculation: Appendix A + Haaf et al. (2020)
rest LLDPE (DSD spec 310)	energy recovery with CO2 capture	own calculation: Appendix A + Haaf et al. (2020)
rest LDPE (DSD spec 310)	energy recovery with CO2 capture	own calculation: Appendix A + Haaf et al. (2020)
rest HDPE (DSD spec 329)	energy recovery with CO2 capture	own calculation: Appendix A + Haaf et al. (2020)
rest PP (DSD spec 324)	energy recovery with CO2 capture	own calculation: Appendix A + Haaf et al. (2020)
rest GPPS (DSD spec 331)	energy recovery with CO2 capture	own calculation: Appendix A + Haaf et al. (2020)
rest HIPS (DSD spec 331)	energy recovery with CO2 capture	own calculation: Appendix A + Haaf et al. (2020)
rest PET (DSD spec 325)	energy recovery with CO2 capture	own calculation: Appendix A + Haaf et al. (2020)
MR residual of LLDPE (DSD spec 310)	energy recovery with CO2 capture	own calculation: Appendix A + Haaf et al. (2020)
MR residual of LDPE (DSD spec 310)	energy recovery with CO2 capture	own calculation: Appendix A + Haaf et al. (2020)
MR residual of HDPE (DSD spec 329)	energy recovery with CO2 capture	own calculation: Appendix A + Haaf et al. (2020)
MR residual of PP (DSD spec 324)	energy recovery with CO2 capture	own calculation: Appendix A + Haaf et al. (2020)
MR residual of GPPS (DSD spec 331)	energy recovery with CO2 capture	own calculation: Appendix A + Haaf et al. (2020)
MR residual of HIPS (DSD spec 331)	energy recovery with CO2 capture	own calculation: Appendix A + Haaf et al. (2020)
MR residual of PET (DSD spec 325)	energy recovery with CO2 capture	own calculation: Appendix A + Haaf et al. (2020)
inert gas	nitrogen as inert gas	conversion of nitrogen flow of ecoinvent without burden
ammonium sulfate	disposal of ammonium sulfate	own calculation: Appendix A
carbon dioxide	CO2 (atm) release	release to atmosphere
carbon dioxide	CO2 (100bar) release	release to atmosphere
carbon dioxide	direct air capture	Deutz and Bardow (2021)
carbon dioxide	CO2 from cement plant (not included in the final assessment)	von der Assen et al. (2016)
carbon dioxide	CO2 from cement plant (not included in the final assessment)	von der Assen et al. (2016)
wood chips	correction of bio CO2 uptake	own calculation: Table 5.1
bark chips	correction of bio CO2 uptake	own calculation: Table 5.1
miscanthus	correction of bio CO2 uptake	own calculation: Table 5.1
wood pellets	correction of bio CO2 uptake	own calculation: Table 5.1

continued on next page

Table C.1 – *continued from previous page*

name of flow	technology	sources
PP packaging waste	landfilling (ecoinvent)	Ecoinvent (2020)
LDPE packaging waste	landfilling (ecoinvent)	Ecoinvent (2020)
LLDPE packaging waste	landfilling (ecoinvent)	Ecoinvent (2020)
HDPE packaging waste	landfilling (ecoinvent)	Ecoinvent (2020)
PET packaging waste	landfilling (ecoinvent)	Ecoinvent (2020)
GPPS packaging waste	landfilling (ecoinvent)	Ecoinvent (2020)
HIPS packaging waste	landfilling (ecoinvent)	Ecoinvent (2020)
PP waste	landfilling (ecoinvent)	Ecoinvent (2020)
LDPE waste	landfilling (ecoinvent)	Ecoinvent (2020)
LLDPE waste	landfilling (ecoinvent)	Ecoinvent (2020)
HDPE waste	landfilling (ecoinvent)	Ecoinvent (2020)
PET waste	landfilling (ecoinvent)	Ecoinvent (2020)
PP&A - PET waste	landfilling (ecoinvent)	Ecoinvent (2020)
GPPS waste	landfilling (ecoinvent)	Ecoinvent (2020)
HIPS waste	landfilling (ecoinvent)	Ecoinvent (2020)
PVC waste	landfilling (ecoinvent)	Ecoinvent (2020)
PUR flex waste	landfilling (ecoinvent)	Ecoinvent (2020)
PUR rigid waste	landfilling (ecoinvent)	Ecoinvent (2020)
PP&A - polyamide 6 waste	landfilling (ecoinvent)	Ecoinvent (2020)
PP&A - polyamide 66 waste	landfilling (ecoinvent)	Ecoinvent (2020)
PP&A - acrylic waste	landfilling (ecoinvent)	Ecoinvent (2020)
acetonitrile	market for acetonitrile (ecoinvent)	Ecoinvent (2020)
ammonium sulfate	market for ammonium sulfate (ecoinvent)	Ecoinvent (2020)
butadiene	market for butadiene (ecoinvent)	Ecoinvent (2020)
butene-1	market for butene, mixed (ecoinvent)	Ecoinvent (2020)
calcium chloride	market for calcium chloride (ecoinvent)	Ecoinvent (2020)
calcium oxide	market for lime (ecoinvent)	Ecoinvent (2020)
caustic soda (50%)	market for sodium hydroxide, without water, in 50% solution state (ecoinvent)	Ecoinvent (2020)
cooling water	market for water, decarbonised, at user (ecoinvent, proxy)	Ecoinvent (2020)
deionized water	market for water, decarbonised, at user (ecoinvent, proxy)	Ecoinvent (2020)
electricity	market group for electricity (ecoinvent)	Ecoinvent (2020)
gasoline	market for petrol, unleaded (ecoinvent)	Ecoinvent (2020)
hydrogen cyanide	market for hydrogen cyanide (ecoinvent)	Ecoinvent (2020)
natural gas (raw material)	market for natural gas, high pressure (ecoinvent)	Ecoinvent (2020)
n-pentane	market for pentane (ecoinvent)	Ecoinvent (2020)
process water	market for water, decarbonised, at user (ecoinvent, proxy)	Ecoinvent (2020)
silicon carbide	market for silicon carbide (ecoinvent)	Ecoinvent (2020)
sodium carbonate	market for sodium bicarbonate (ecoinvent)	Ecoinvent (2020)

continued on next page

Table C.1 – *continued from previous page*

name of flow	technology	sources
sulfur trioxide	market for sulfur trioxide (ecoinvent)	Ecoinvent (2020)
sulfuric acid	market for sulfuric acid (ecoinvent)	Ecoinvent (2020)
vegetable oil	market for vegetable oil methyl ester (ecoinvent)	Ecoinvent (2020)
ethane	ethane extraction, from natural gas liquids (ecoinvent)	Ecoinvent (2020)
wood chips supply	market for wood chips, wet, measured as dry mass (ecoinvent)	Ecoinvent (2020)
bark chips supply	market for bark chips, wet, measured as dry mass (ecoinvent)	Ecoinvent (2020)
miscanthus supply	market for miscanthus, chopped (ecoinvent)	Ecoinvent (2020)
wood pellets supply	wood pellet production (ecoinvent)	Ecoinvent (2020)
hydrochloric acid	market for hydrochloric acid, without water, in 30% solution state (ecoinvent)	Ecoinvent (2020)
monoethanolamine	market for monoethanolamine (ecoinvent)	Ecoinvent (2020)
naphtha	market for naphtha (ecoinvent)	Ecoinvent (2020)

C.2 Direct electrification technologies

To directly electrify chemical processes, steam production and electrical furnaces can be used. These electric furnaces (efurnaces) directly utilize renewable electricity in resistive (ohmic) (Rieks et al., 2015; Wismann et al., 2019) or inductive heaters (Vinum et al., 2018) and thereby enable fully electrical driven steam reformers for the large-scale production of hydrogen, syngas, or basic chemicals, such as ethylene and propylene. For the electrified steam boiler, data from the IHS Process Economics Program is used (IHS Markit, 2018).

The resistive heating unit-process includes only the amount of required silicon carbide conductor material. The required quantity of 2.58 10e-7 kg SiC per MJ thermal energy is based on simplified assumptions: a radiant heater with a power of 100 kW and a temperature of 1200°C is assumed. With 8000 hours per year and ten years of operation, 28.8 TJ are supplied over the lifetime. The emission ratio of SiC at 1200 °C is 0.92 (Brandt et al., 1998). According to the Stefan-Bolzmann Law, the required radiator area results in 0.924 m2. A radius of 5 mm, assuming a cylindrical heating conductor with negligibly small end areas (length much larger than the radius) and a density of 3,217 kg per m3 of silicon carbide, gives 7.43 kg of silicon carbide.

C.3 Renewable ammonia production

The Haber-Bosch process produces ammonia: a gaseous mixture of $H_2$and $N_2$is sent to a reactor operating at approx. 450°C and 200bar. The subsequent mixture is cooled such that liquid ammonia can be stored cryogenically. The remaining hydrogen and nitrogen are sent back to the reactor (Elvers and Ullmann, 2011). In the analysis, unit process data according to Cesaro et al. (2021) is used. An air separation unit supplies nitrogen. Hydrogen must be supplied by water electrolysis or other means.

C.4 Water electrolysis

The data for the electrolysis of water to hydrogen and oxygen is based on a low-temperature process. This process assumes an average efficiency of 67.15% and, thereby, consumes 178.7 MJ per kg of hydrogen (Agora, 2018).

C.5 Biomass gasification

The gasification of lignocellulosic biomass to syngas is based on the pressurized direct oxygen-steam blown circulating fluidized bed gasifier (CFB) (Arvidsson et al., 2014; Hannula and Kurkela, 2010, 2012; Isaksson et al., 2012; Spath et al., 2005) and an atmospheric indirect air-blown dual fluidized bed gasifier (DFB) (Arvidsson, M. et al., 2012). As biomass feed, wood chips, wood pellets, bark chips, and miscanthus are considered. Both gasifiers were adapted to the actual feedstock composition and ash contents. The output synthesis gas has a hydrogen to carbon monoxide ratio of 2:1. CO_2 from synthesis gas upgrading is captured and can be used as feedstock. Excess heat is provided in the form of steam.

C.6 Biomass fermentation

The ethanol production from lignocellulosic biomass is based on the enzymatic saccharification and co-fermentation of generated sugars (Humbird et al., 2011) from wood chips, wood pellets, and miscanthus. Bark chips are excluded because no inventory could be obtained. In the first step, lignocellulosic biomass is pre-treated with diluted sulfuric acid to free hemicellulose sugars and break down the biomass.

Ammonia is added to raise pH levels. Afterward, the slurry is infused with a cellulase enzyme produced on-site to split the cellulose into fermentable sugars. The microorganism Zymomonas mobilis is added to ferment the sugars into ethanol. The resulting beer is sent to the purification section, where solids are separated, and the ethanol is distilled close to azeotrope. The remaining water is separated from the ethanol using vapor-phase molecular sieve adsorption yielding 99.5% pure ethanol. Solids from the separation and the wastewater treatment and biogas from wastewater treatment are combusted to supply electricity and process heat. The boiler produces excess steam that can be used to produce electricity for the grid or heat for close-by processes. For each lignocellulose feedstock, the composition of, e.g., cellulose, hemicellulose, lignin, and ash is adapted. Additionally, steam is exported from the high and medium-pressure turbine as a by-product.

C.7 Biomass for steam production

Lignocellulosic steam boilers are included since they have been shown to provide an economical alternative to natural gas boilers (Pérez-Uresti et al., 2019). Producing steam from biomass is based on the steam boiler's energy demand from the IHS Process Economics Program (IHS Markit, 2018). The energy content of the biomass supply is divided by the energy demand of the steam boiler or approx. 3.4 MJ per kg of steam.

C.8 Chemical recycling - pyrolysis

The production of liquid hydrocarbons from plastic waste that can be used in steam crackers is based on pyrolysis. The process is based on industrially verified data (IHS Markit, 2018) and own modeling. It includes feed preparation and dehydrochlorination in order to be able to handle the plastic waste of various compositions. Magnetic fields extract metal parts. Plastic waste input is subsequently broken down into smaller pieces. In the pyrolysis reactor, plastic waste is heated such that liquid hydrocarbons are formed. The overall carbon efficiency is assumed to be 63% while energy demands depend on the amount of liquid hydrocarbon output and vary between 0.17 kg and 0.44 kg per kg of plastic packaging waste input. Additionally, caustic soda (50%) is needed to treat off-gases. Depending on the plastic waste, the amount of caustic soda (50%) varies between 0.12 kg to 0.32 kg per kg plastic waste input. Resulting waste is treated in hazardous waste incinerators.

Appendix C Documentation of data sources and additional results

C.9 Sankey diagram of circular-carbon pathway

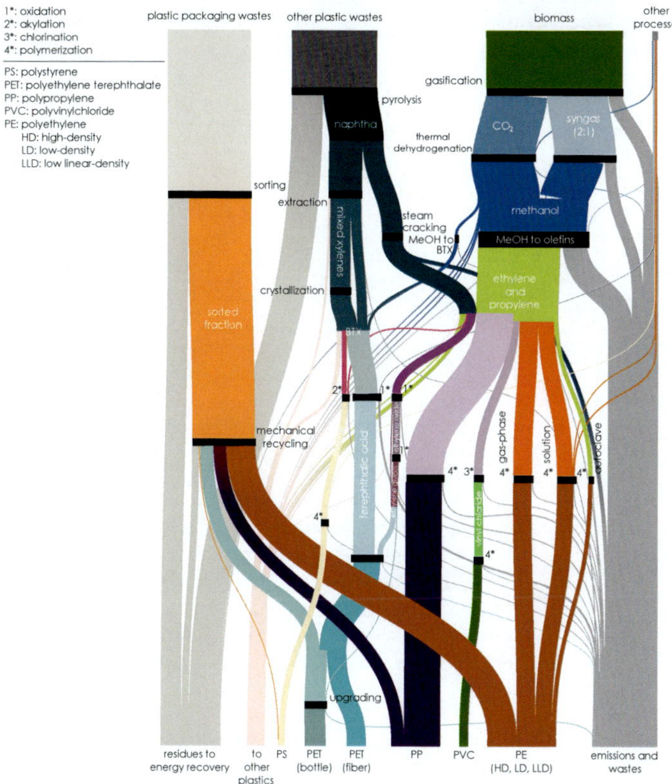

Figure C.1: The width of each flow represents the relative carbon content of the flow. Each black bar represents a chemical production process converting input to output flows. Since the production volumes of polyamide 6, polyamide 66, polyacrylonitrile fibers, flexible polyurethanes, and rigid polyurethanes are each small compared to the other plastics, their supply chain is not fully resolved and is labeled "to other plastics" for simplicity. To further increase readability, some minor chemical production technologies are grouped as "other processes". The emissions and wastes include all carbon-containing residues directly emitted (e.g., process-related CO_2 emissions) or sent to waste incineration (e.g., residual wastes from chemical production facilities).

C.10 Additoinal results including early-stage technologies

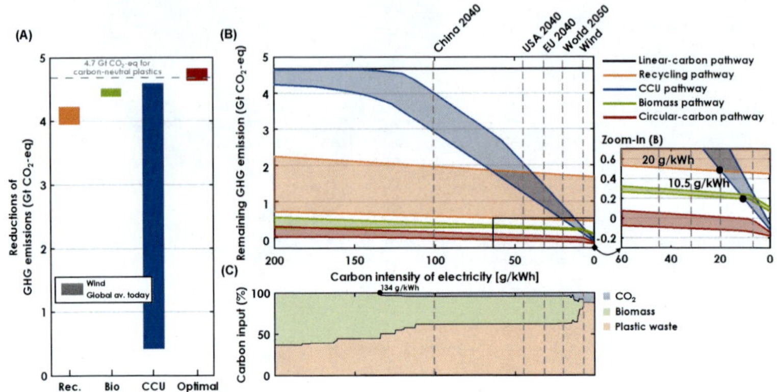

Figure C.2: (A) Reductions of greenhouse gas emission in Gt of CO2-eq of the four pathways: recycling, biomass utilization, Carbon Capture and Utilization (CCU), and the optimal combination thereof (circular-carbon pathway). (B) The remaining greenhouse gas emissions in Gt CO2-eq of the linear-carbon pathways and the four circular pathways depending on the carbon intensity of electricity in gram CO2-equivalent per kWh. (C) Optimal carbon input in % of the circular carbon pathway depending on the carbon intensity of electricity in gram CO2-equivalent per kWh. The IEA report (61) only provides regionalized electricity impacts until 2040.

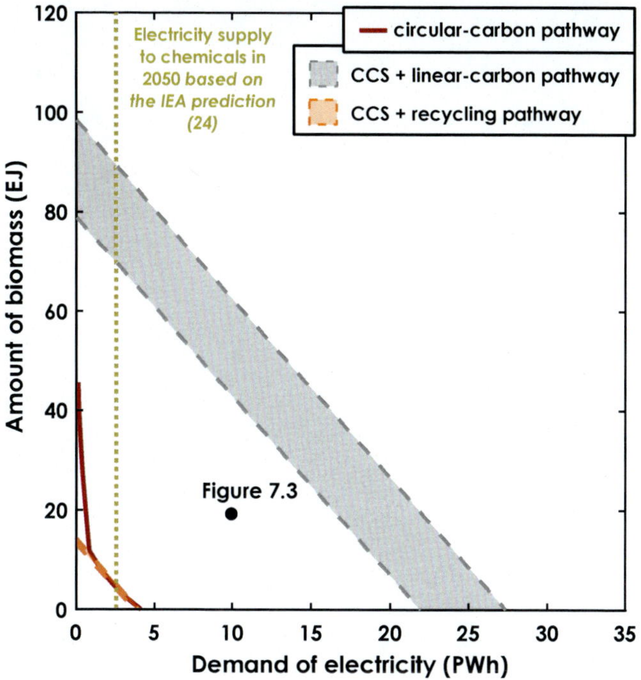

Figure C.3: The amount of biomass and electricity for the circular-carbon pathway is shown. Additionally, the fossil industry and the recycling scenario together with CCS are shown. The range of CCS reflects different sources of CO2, such as ammonia production plant (62) or ambient air (63). The energy demands for the linear-carbon and maximal recycling scenario are based on fossil resources, and only the equal amount of biomass and/or electricity is depicted here.

C.11 Price ranges

Table C.2: Price ranges for each resource supply and the waste treatment (IEA, 2020; IHS Markit, 2018; Rudolph et al., 2017)

Resource / Waste volume	Oil	Biomass	CO_2	Electricity	Mech. recycling	Chem. recycling	Energy recovery
Unit	GJ oil-eq	GJ	t	MWh	t	t	t
Lower price / in USD per unit	8.78	5	30	20	200	510	140
Higher price / in USD per unit	12.5	15	142	60	200	510	140

Appendix D

Reducing greenhouse gas emissions by a CO_2-based chemical industry

CO_2 can be utilized to produce several chemicals (Centi et al., 2013; Kondratenko et al., 2013). Making these CO_2-based chemicals is seen as promising to remove CO_2 emissions from the atmosphere and avoid fossil-based chemical production. Thereby, CO_2-based chemicals are advocated to reduce greenhouse gas emissions (Artz et al., 2018). These potential reductions of greenhouse gas emissions have been reported for individual CO_2-based chemicals, while the global reduction potential for greenhouse gas emissions is currently estimated based on the amount of stored CO_2 (Mac Dowell et al., 2017). However, the potential to reduce greenhouse gas emissions per kg of stored CO_2 has been shown to be as high as 12 kg CO_2-eq per kg CO_2 used (cf. Chapter 4). Thus, the comprehensive evaluation of the potential to reduce greenhouse gas emissions is missing. In this appendix, the technical potential to reduce greenhouse gas emissions by a CO_2-based chemical industry is evaluated.

Major parts of this chapter are reproduced with permission of the Proceedings of the National Academy of Science of the United States of America from:

Kätelhön, A., Meys, R., Deutz, S., Suh, S. and Bardow, A. (2019). Climate change mitigation potential of carbon capture and utilization in the chemical industry. *Proceedings of the National Academy of Sciences of the United States of America*, 116(23): 11187-11194.

The author of this thesis contributed to the publication by providing the underlying model and including the respective dataset for the fossil-based processes and the carbon capture and utilization technologies. Furthermore, the author of this thesis contributed to designing the research question, performing the research, and writing the initial draft of the paper.

D.1 Scope of the analysis in relation to this thesis

This appendix focuses on bulk chemicals and provides an in-depth analysis of carbon capture and utilization technologies. However, to calculate the results, the same computational structure and bottom-up model as for plastics is used (cf. Chapter 5).

The major difference is that the production volumes in Table D.1 are used instead and the following 20 major bulk chemicals are considered: acrylonitrile, ammonia, benzene, caprolactam, cumene, ethylene, ethylene glycol, ethylene oxide, methanol, mixed xylenes, phenol, polyethylene, polypropylene, propylene, propylene oxide, p-xylene, styrene, terephthalic acid, toluene, and vinyl chloride. As a result, the scope of the analysis is shifted from plastics to bulk chemicals. In particular, the overall amount of produced high-volume chemicals is higher since also other uses than plastics are fully covered.

Furthermore, as end-of-life scenario, the incineration of all chemicals in a waste incinerator is assumed, leading to 2.2 Gt CO_2-eq emissions. The assumption of waste incineration is a worst-case assumption for end-of-life emissions. Assuming waste incineration neglects that parts of the chemicals may also be stored in materials for substantially long times. Both the use phase and the end of life of chemicals are assumed to be identical because all CO_2-based technologies considered produce identical chemicals as conventional production. Therefore, potential benefits from temporary carbon storage or sequestration will be the same in the conventional and the CO_2-based scenarios and not affect the difference in greenhouse gas emissions.

D.2 Scenarios for a CO_2-based chemical industry

The bottom-up model of the chemical industry yields future production pathways for the production of 20 large-volume chemicals in 2030 (Ausfelder et al., 2013) for three scenarios. In the "conventional scenario" (cf. Figure D.1), the chemical industry converts 1.12 Gt of fossil resources into chemicals and generates about 2.0 Gt CO_2-eq cradle-to-gate emissions using conventional production technologies that are already commercialized today.

In a second scenario, where the production of the 20 chemicals relies on CO_2-based methanol (Rihko-Struckmann et al., 2010) and methane (Rönsch et al., 2016), feedstock mass flows increase substantially by 287% due to the CO_2 used and water co-produced (cf. Figure D.2): 3.72 Gt CO_2 and 0.59 Gt hydrogen from water electrolysis are annually converted into methanol and methane. Methanol is further processed into

Table D.1: Table of production volumes and final demands in 2030 for the 20 assessed bulk chemicals (Ausfelder et al., 2013; Bazzanella and Ausfelder, 2017).

product	production volume [Mt per year]	final demand [Mt per year]
acetone	10.68	10.68
acrylonitrile	10.42	10.42
ammonia	229.17	220.12
benzene	65.97	11.06
caprolactam	3.48	3.48
cumene	24.32	0.91
diethylene glycol	4.4	4.4
ethylene	250.01	10.55
ethylene glycol	41.66	41.66
ethylene oxide	41.66	10.45
methanol	135.42	135.42
phenol	17.35	17.35
polyethylene	163.19	163.19
polypropylene	100.71	100.71
propylene	139.18	2.93
propylene oxide	13.89	13.89
styrene	48.61	48.61
terephthalic acid	100.69	100.69
toluene	34.7	31.29
vinyl chloride	55.56	55.56
mixed xylenes	93.74	3.69
para-xylene	69.45	1.36

all high-volume chemicals (ethylene, propylene, benzene, toluene and xylene), which are the basis of the other large-volume chemicals. Methane substitutes natural gas to produce both heat and ammonia. CO_2 can be captured from two different sources representing the upper and lower bound for the greenhouse gas emissions of CO_2 supply: a highly concentrated industrial point source (100% CO_2) and ambient air with about 400 ppm CO_2 (von der Assen et al., 2016). This scenario leads basically to the methanol economy envisioned by Asinger (Asinger, 1986) and promoted by Olah (Olah et al., 2009). Since all technologies in this scenario are at technology readiness levels (TRLs) of 7 and higher, this scenario is denoted "high-TRL scenario". The integration of CO_2 via methane and methanol would require comparably low research and development efforts and allow use of wide parts of the existing petrochemical infrastructure and technologies.

The third scenario further includes CO_2-based technologies for the direct conversion of CO_2 into ethylene and propylene (Yabe et al., 2017; Gao et al., 2017), benzene, toluene and xylene (Wang et al., 2014a), carbon monoxide (Artz et al., 2018), ethylene oxide (Mobley et al., 2017), and styrene (Wang et al., 2014b) (cf. Figure D.3). These technologies are currently at early research and development stages with TRLs below 7. Therefore, this scenario is denoted "low-TRL scenario." The direct synthesis of chemicals from CO_2 reduces the amount of CO_2 and hydrogen needed to 2.77 and 0.38 Gt, respectively, but mass flows are still about 182% larger compared with the conventional scenario. Due to the early development stages of these technologies, the realization of this scenario would require substantial research and development efforts and novel production facilities. Since this scenario is based on low-TRL technologies, all calculations are subject to increased data uncertainty.

In the CO_2-based scenarios, hydrogen produced via water electrolysis represents the major source of energy for the chemical industry. Electricity demands amount to about 32.0 PWh in the high-TRL scenario and 18.1 PWh in the low-TRL scenario. Electricity demands could be reduced through the direct use of hydrogen from water electrolysis. For example, the direct use of hydrogen in ammonia production could reduce the electricity demand in the low-TRL scenario by 7.5% by avoiding the steps for methanation and steam reforming in the production of ammonia from CO_2-based methane. However, these production pathways are not based on CO_2 and therefore excluded from the CO_2-based scenarios. For the CO_2-based scenarios, the electricity demands in the high- and low-TRL scenario correspond to about 9 % and 55% of the expected world electricity production in 2030 according to the new policy scenario of the International Energy Agency (IEA, 2018b), respectively. Hence, CO_2 utilization in the chemical industry could only reduce greenhouse gas emissions on a large scale with the massive expansion of electricity production capacities. As a result, the carbon footprint of electricity from the technologies used to expand the electricity production capacities will determine the benefits of CO_2-based chemical production for greenhouse gas emissions.

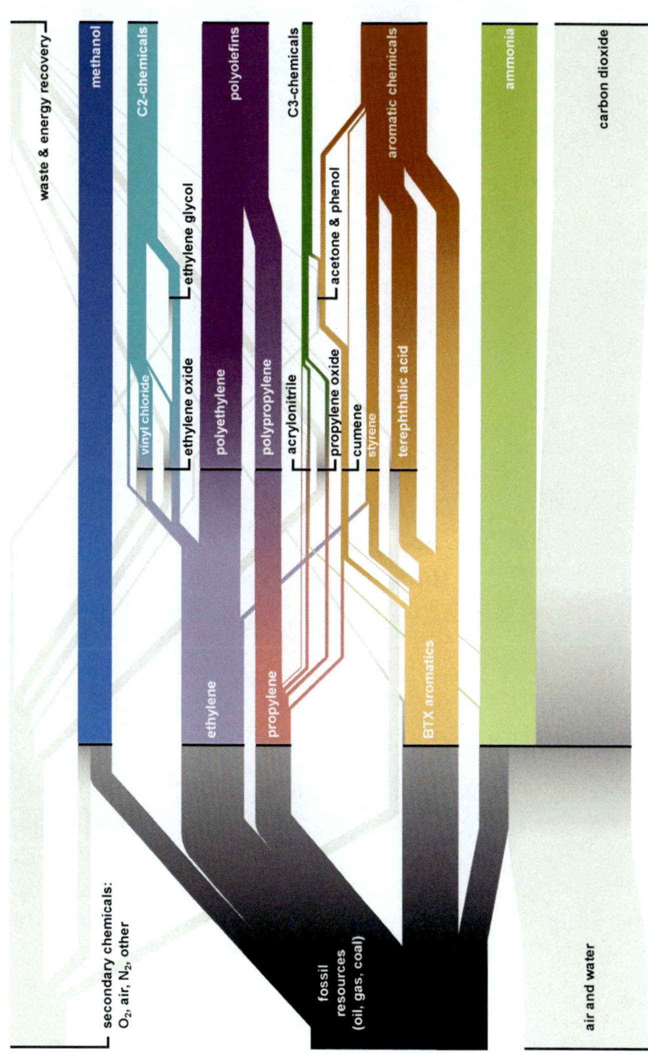

Figure D.1: Mass flows within the chemical industry in the conventional scenario. The projected final demand for 20 large-volume chemicals in 2030 is produced.

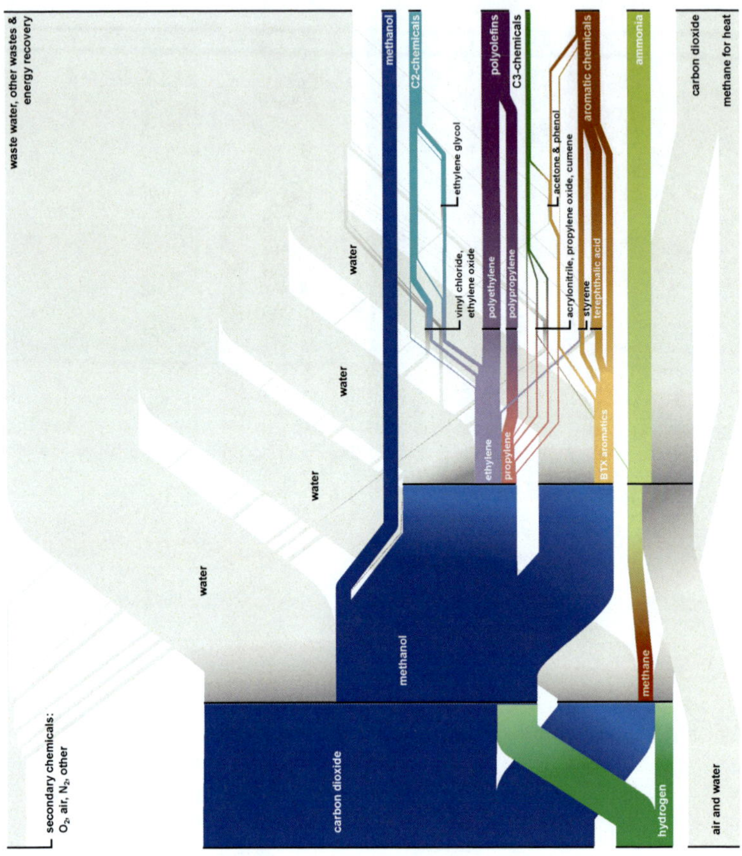

Figure D.2: Mass flows within the chemical industry in the high-TRL scenario. The projected final demand for 20 large-volume chemicals in 2030 is produced.

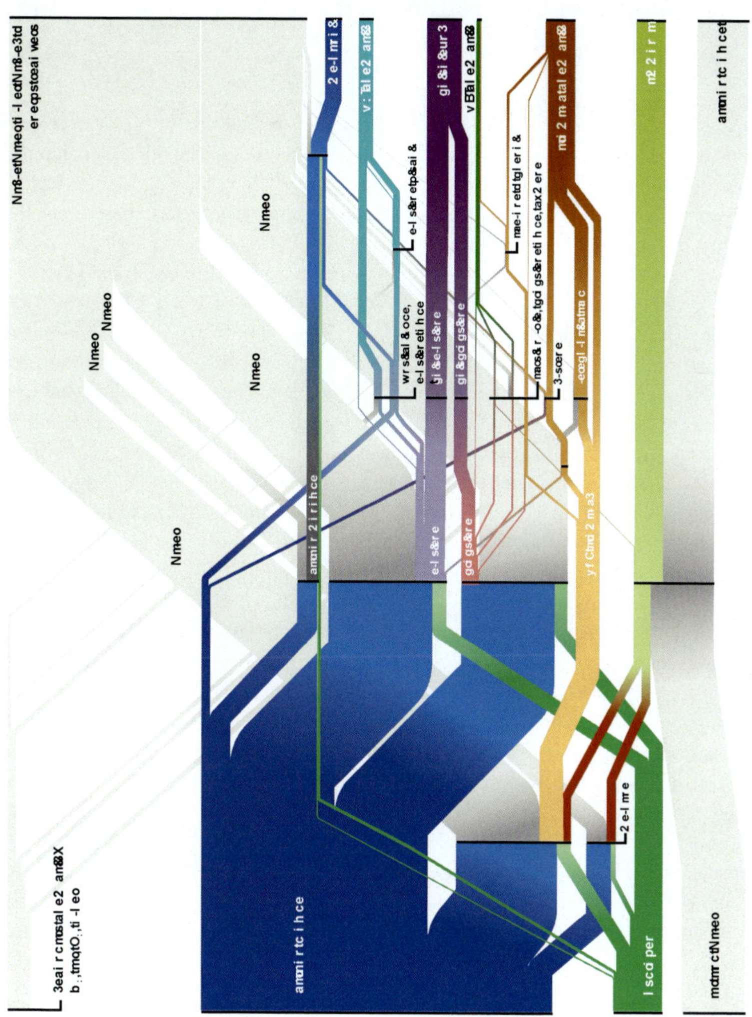

Figure D.3: Mass flows within the chemical industry in the low-TRL scenario. The projected final demand for 20 large-volume chemicals in 2030 is produced.

D.3 Reductions of greenhouse gas emissions depending on electricity supply

The results show that the greenhouse gas emission reductions of replacing conventional chemical production through CO_2-based production vary widely depending on the carbon footprint of electricity (cf. Figure D.4). The results show that utilizing CO_2 can substantially reduce greenhouse gas emissions in both CO_2-based scenarios, but only if low-carbon electricity is available. In the high-TRL scenario, CO_2-based technologies reduce greenhouse gas emissions compared with conventional technologies once the employed electricity has a carbon footprint of 260 g CO_2-eq per kWh or lower. The greenhouse gas emissions of the chemical industry decline non-linearly as a function of the carbon footprint of electricity since the optimization approach minimizes greenhouse gas emissions and different CO_2-based technologies start to out-compete the corresponding conventional technologies in terms of their greenhouse gas emissions at different carbon footprints of electricity. Electricity carbon footprints where CO_2-based technologies enter the production mix in the model range from 74 g CO_2-eq per kWh (mixed xylene) to 260 g CO_2-eq per kWh (methanol). For all technologies, capturing CO_2 from the industrial point source leads to higher emission reductions than air capture because less energy is required for capture (von der Assen et al., 2016). Greenhouse gas emissions are most reduced by using CO_2 from the industrial point source and electricity from wind power because wind power is currently the energy generation technology with the lowest carbon footprint. The resulting maximum greenhouse gas emission reduction is the difference between the greenhouse gas emissions of the conventional and of the CO_2-based scenario. The life cycles of conventional and CO_2-based technologies differ only from cradle-to-gate, while emissions during the use phase and end of life, as well as potential benefits from carbon storage are identical in all scenarios since the same chemicals are produced. The maximum greenhouse gas emission reductions amount to 3.4 Gt CO_2-eq per year and represent the technical greenhouse gas mitigation potential of the scenario. Emissions are reduced by replacing emission-intensive processes within the chemical industry and its supply chain, and by capturing CO_2 outside the boundaries of the chemical industry. The resulting structure of the chemical industry is presented in Figure D.2.

In the low-TRL scenario, moderate greenhouse gas emission reductions are achieved regardless of the carbon footprint of electricity, because this scenario also includes CO_2-based technologies that do not require electricity for hydrogen production, i.e., synthesis of ethylene from CO_2 and methane, and ethylene oxide from CO_2 and ethylene. However, also in the low-TRL scenario, emissions can be substantially reduced only with low-carbon electricity. CO_2-based technologies become beneficial for electricity carbon footprints between 44 g CO_2-eq per kWh (carbon monoxide) and 334 g CO_2-eq per kWh (para-xylene). In the low-TRL scenario, the annual emissions are reduced from 4.2 Gt CO_2-eq for the conventional scenario to 0.7 Gt CO_2-eq leading to a maximum potential of 3.5 Gt CO_2-eq per year.

To contextualize the required carbon footprints of electricity, the current carbon footprint of grid electricity for different countries and technologies in Figure D.4 is shown. While most countries and technologies do not provide electricity with a sufficiently low carbon footprint to realize substantial greenhouse gas emission reductions by CO_2-based chemicals, there are countries where CO_2-based chemicals could reduce greenhouse gas emissions substantially using grid electricity already today. In France, for example, the introduction of high-TRL CO_2-based technologies would roughly halve the greenhouse gas emissions of chemicals over the entire life cycle.

Even though individual countries have already demonstrated that low carbon electricity can be provided at scale, the transformation of the chemical industry toward CO_2 utilization would cause a substantial demand for additional electricity that would require a large-scale expansion of electricity production capacities. This expansion would require large shares of renewable energy to achieve a sufficiently low carbon footprint.

If all additional electricity were provided by renewable energy, the amount of renewable energy required for the full-scale introduction of CO_2-based chemicals would correspond to 126% and 222% of current targets (sustainable development scenario of IEA (IEA, 2018b)) for the global renewable electricity production in 2030 for the low-TRL and the high-TRL scenario, respectively. Even these current targets are already known not to be achievable based on existing and announced policies (IEA, 2018b). Thus, the need for a further expansion of renewable electricity production capacities is likely to be a limiting factor for CO_2 utilization in the chemical industry.

The potential to reduce greenhouse gas emissions of CO_2-based technologies depends on both the amount of additional electricity available and the carbon footprint of electricity (cf. Figure D.5). The amount of additional electricity available determines the scale of CO_2 utilization and resulting greenhouse gas emission reductions. For a given electricity carbon footprint, the greenhouse gas emissions of the chemical industry depend non-linearly on the amount of electricity available, because individual CO_2-based technologies differ in their efficiencies in using renewable energy to achieve emission reductions. The model allocates additional electricity to the CO_2-based technologies that save the largest amount of greenhouse gas emissions until the demands for the chemicals from such technologies are exhausted. Additional electricity is, thus,

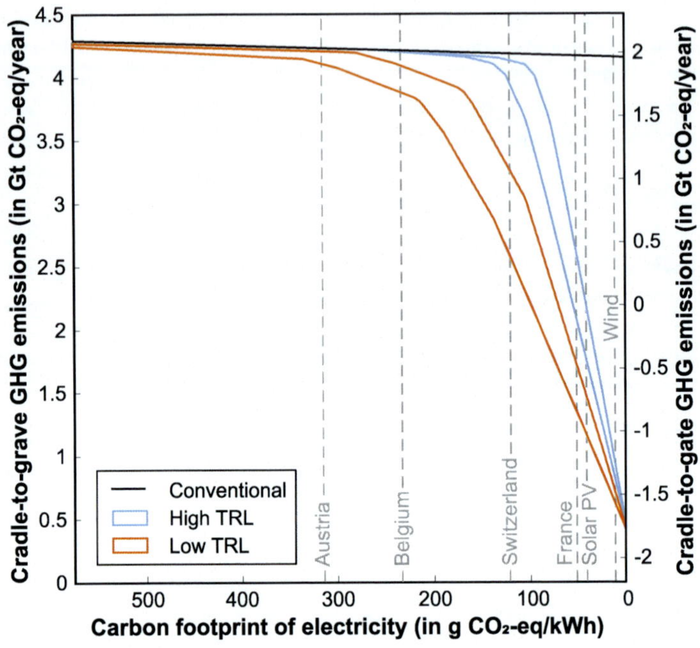

Figure D.4: Cradle-to-grave and cradle-to-gate greenhouse gas emissions of the chemical industry producing the final demand for 20 large-volume chemicals in 2030 as function of the carbon footprint of electricity. Cradle-to-grave emissions include all emissions throughout the life cycles of the 20 large-volume chemicals, while cradle-to-gate emissions cover only the production stage and the supply of all raw materials and energy needed for production. The vertical dashed lines illustrate the greenhouse gas emissions of grid electricity in selected countries or of electricity from selected renewable energy technologies.

progressively allocated to CO_2-based technologies reducing greenhouse gas emissions less efficiently. As a result, the slope of the greenhouse gas emissions decreases with the amount of additional electricity available. For carbon-free electricity, efficiencies of CO_2-based technologies range from 340 g CO_2-eq avoided per kilowatt-hour for the production of para-xylene to 140 g CO_2-eq avoided per kilowatt-hour for the production of methane. Hence, higher emission reductions per kilowatt-hour can be achieved, if the limited low-carbon electricity is used for the most efficient CO_2-based technologies.

D.4 Discussion of results

Considering both the large potential of CO_2 utilization in the chemical industry to mitigate greenhouse gas emissions and the large amount of low-carbon electricity needed to exploit this potential, two questions arise: (i) Can we provide that much low-carbon electricity, and, if yes, (ii) should we use it in the chemical industry?

According to a recent Intergovernmental Panel on Climate Change (IPCC) report (Edenhofer, 2012), the amount of renewable energy that could be produced by full implementation of demonstrated technologies exceeds even the most ambitious scenarios for renewable energy deployment in 2050 by more than one order of magnitude. Thus, in theory, it is possible to produce sufficient low-carbon electricity. However, the actual implementation of renewable energy capacities is limited by practical constraints like costs, competition for land, public acceptance of energy infrastructure, and limits in the uptake of intermittent electricity supply by the electricity grid. Despite the remarkable technological progress and substantial reduction in cost, modern non-hydro renewable power, including wind, solar, biomass, geothermal, and ocean power, accounts for only 1.7% of the global final energy demand (REN21, 2018). Hence, dramatically increasing low-carbon electricity to the extent that it can sufficiently power CO_2 utilization for chemicals in the next decade is likely to be a challenge, and renewable energy is expected to remain a limited resource for the coming decades.

In this case, how much of the limited, low-carbon electricity resource should be allocated to CO_2-based technologies compared with other potential uses for low-carbon electricity? For this question, it is useful to differentiate between on- and off-grid use of renewable electricity. For on-grid use, the additional low-carbon electricity for CO_2 utilization is distributed via the electricity grid. In this case, the electricity could also be used by other greenhouse gas mitigation technologies connected to the grid such as e-mobility or heat pumps. From the point of reducing greenhouse gas emissions, limited low-carbon electricity should be used for those technologies that reduce the greenhouse gas emissions the most (Schakel et al., 2017). In Figure D.5, as an example, the greenhouse gas emission reduction by substituting natural gas boilers with either heat pumps or electric boilers, as well as substituting diesel or gasoline cars through e-mobility are shown. Power-to-heat and power-to-e-mobility are both very large-scale uses of renewable electricity and would, thus, compete with CO_2 utilization for electricity. The greenhouse gas emission reductions have been calculated based on

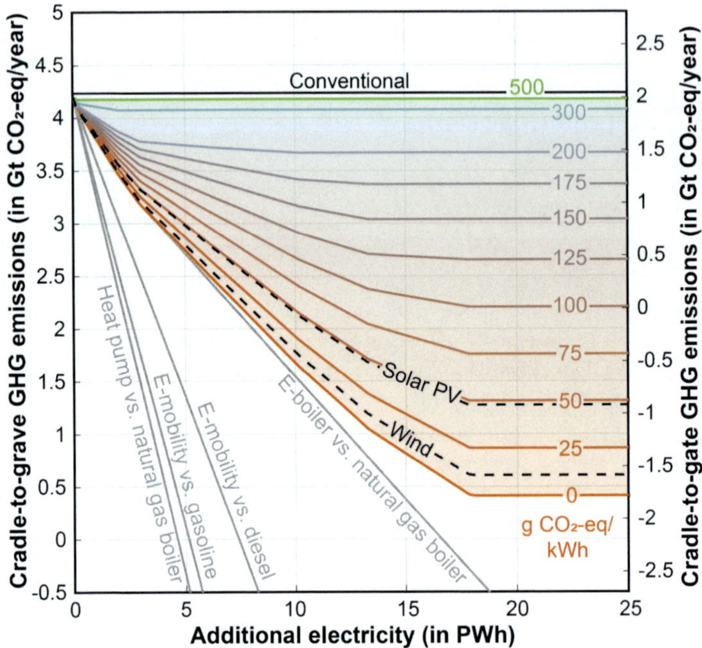

Figure D.5: Cradle-to-grave and cradle-to-gate greenhouse gas emissions for the chemical industry in the low-TRL scenario producing the final demand for 20 large-volume chemicals in 2030 as function of the amount of additional electricity available and its carbon footprint in grams of CO_2 equivalent per kilowatt hour. Cradle-to-grave emissions include all emissions throughout the life cycles of the 20 large-volume chemicals, while cradle-to-gate emissions cover only the production stage and the supply of all raw materials and energy needed for production. The black solid line represents the greenhouse gas emissions in the conventional scenario. The light gray lines illustrate potential greenhouse gas emission reductions by using the additional electricity for e-mobility to substitute gasoline or diesel cars, for heat generation in a heat pump, or an e-boiler to substitute heat from natural gas boilers from Sternberg and Bardow (2015) (cf. Section D.5.

power-to-X efficiencies by Sternberg and Bardow (Sternberg and Bardow, 2015) for the use of emission-free electricity from surplus power (cf. Section D.5). The power-to-X efficiency represents the greenhouse gas emission reductions per amount of electricity used. In comparison with all hydrogen- and CO_2-based technologies considered in this thesis, e-mobility and heat pumps reduce the greenhouse gas emissions more strongly per kilowatt-hour of electricity used. Hence, from a perspective of greenhouse gas emissions, the implementation of these technologies should be prioritized over the hydrogen- and CO_2-based technologies considered in this thesis until their demand for renewable energy is fully exhausted. However, hydrogen and CO_2 utilization in the chemical industry still provides a later option to reduce greenhouse gas emissions, since large-scale emission reductions will eventually be needed in all relevant sectors (Edenhofer, 2012). Similarly, even the installation of electric boilers to substitute natural gas boilers reduces the greenhouse gas emissions more than all hydrogen- and CO_2-based technologies except the CO_2-based production of para-xylene and styrene. Still, emission benefits are achieved by CO_2-based technologies that do not require electricity for hydrogen production such as the production of ethylene and ethylene oxide.

However, the perspective changes in remote areas without grid connection but with large unused renewable energy resources. Here, off-grid CO_2-based technologies could provide benefits without competing with other technologies for renewable energy. Even more, by transforming regionally bound renewable energy into transportable commodities, CO_2-based technologies could enable global trade of renewable energy, for example, in the form of CO_2-based methanol or methane. Also, ammonia could be used as renewable fuel (Cédric Philibert, 2017b). However, ammonia is produced more efficiently by the direct conversion of renewable hydrogen with nitrogen than by CO_2-based methane reforming. Setting up production plants in remote areas will result in additional costs for infrastructure, feedstock supply (e.g., CO_2 and water), and the transportation of the CO_2-based product (e.g., methanol) to established chemical production sites for further processing. For the CO_2-based production of methanol in remote areas, for example, additional transportation costs for the CO_2 supply from industrial point sources and the transportation of methanol have been estimated to about 100 USD/t, i.e., about 25% of current methanol production costs based on fossil resources (Cédric Philibert, 2017b). These additional costs, however, will partly be offset by lower renewable electricity prices and higher load factors in the world's most favorable production regions (Cédric Philibert, 2017b). Using hourly wind and solar geospatial data, Fasihi et al. (2016) identified wide areas where a combination of one axis sun-tracking solar photovoltaic capacities and modern wind turbines could enable water electrolysis with load factors higher than 50%. Many of these areas are located in Africa, Australia, and South America, where the amount of available renewable energy resources is more than 50 times higher than the current total primary energy demand (Edenhofer, 2012). Even though the study does not quantify potential benefits from using CO_2-based energy vectors outside the chemical industry, the trade of renewable energy through CO_2-based products could be highly beneficial, especially, for densely populated regions such as Germany, Japan, or Korea that have fewer renewable energy resources and may in practice fall short of employing sufficient capacities to satisfy their future renewable energy demand (Edenhofer, 2012).

The presented analysis focuses on the technical potential to reduce greenhouse gas emissions by CO_2 utilization in the chemical industry in 2030, i.e., the maximum reduction of greenhouse gas emissions that is technically feasible in case of full deployment of CO_2 utilization. The actual deployment rate, however, is likely to be lower. First, achieving full deployment of novel CO_2-based technologies by 2030 would imply an average annual increase in market penetration of 10% between 2020 and 2030. This increase in market penetration substantially exceeds current replacement rates of chemical production plants: Assuming an average plant life time of 25 years (39), only 4% of existing production capacities are replaced annually. Consequently, even though parts of the existing chemical infrastructure could still be used, a rapid implementation of CO_2-based technologies would require substantially higher investment volumes, as well as the phase-out of existing production plants before the end of their technically feasible lifetime. Second, the full implementation of CO_2-based technologies can only be achieved with a massive expansion of low-carbon electricity production capacities that substantially exceed current targets for low-carbon electricity production for 2030 (IEA, 2018b). This expansion is likely to represent a bottleneck for large-scale hydrogen and CO_2 utilization within the next decades. Furthermore, the expansion will be based on a mix of electricity generation technologies leading to a higher carbon footprint than wind power assumed for the technical potential in this study. Third, in addition to CO_2 utilization, other measures have been proposed for reducing greenhouse gas emissions in the chemical industry including the use of biomass feedstock (Spierling et al., 2018), electrification (Wismann et al., 2019), and carbon capture and storage (Mac Dowell et al., 2017). Thus, greenhouse gas emissions of the chemical industry are likely to be reduced by a mix of measures.

The technical potential of CO_2 utilization to reduce greenhouse gas emissions is affected by the CO_2 source. Capturing CO_2 from highly concentrated CO_2 sources such as industrial point sources reduces emissions more, because less energy is required for capture than for dilute sources such as air. The emissions reductions capture how much emissions are changing as consequence of introducing CO_2-based technologies.

However, for the goal of an overall net-zero emission society, the industrial point source itself would have to be de-fossilized or end-of-life emissions from the CO_2-based product would need to be avoided, e.g., by permanent carbon storage. In contrast, direct air capture and biogenic point sources (e.g., bio-based ethylene production) would allow for a closed carbon cycle even if the CO_2-based product is incinerated. Switching toward lower concentrated sources will increase costs due to higher energy demands for CO_2 capture. In a net-zero emission economy, CO_2 utilization could then provide the carbon feedstock needed for the production of a wide range of products (e.g., chemicals, plastics, and pharmaceuticals).

While a detailed economic assessment of CO_2 utilization in the chemical industry is currently not possible due to the early development stage of most CO_2-based technologies, the magnitude of additional operational cost can be estimated based on major cost factors including the avoided annual consumption of crude oil (4.21 billion barrel) and natural gas (6.59 PWh) and the production of hydrogen from low-carbon electricity in the CO_2-based scenarios (0.38 Gt and 0.59 Gt in the low- and high-TRL scenario). With expected prices in 2030 of 92.3 USD per barrel oil (current policy scenario (IEA, 2018b)), 0.017 USD per kWh natural gas (Statistica, 2018), and 1.8-3.5 USD per kg hydrogen (Cédric Philibert, 2017a) from renewable energy in most beneficial production regions (lower limit) and regions with good but not excellent solar and wind resources (upper limit; Europe or similar), the additional operating cost for the full implementation of CO_2-based technologies in the low- and high-TRL scenario amount to 185-833 and 564-1,570 billion USD annually, respectively. These additional costs correspond to 19-87% and 59-164% of the market value of the produced chemicals according to 2017 prices (IHS Markit, 2018). The corresponding costs range between 52-235 USD per t CO_2-eq and 168-467 USD per t CO_2-eq in the low- and high-TRL scenario. For the low-TRL scenario, the lower bound of the costs from this simplified calculation are, thus, in the range of expected carbon prices for the year 2030 (Edenhofer, 2012). This estimate, however, neglects all required investment costs within the chemical industry except for hydrogen production by water electrolysis. In the CO_2-based scenarios, investment costs are likely to increase, because larger mass flows need to be moved: The mass flow of hydrogen and CO_2 needed is 1.8 and 2.9 times larger than the mass flow of fossil inputs in the low- and high-TRL scenario, respectively. The larger mass flows will lead to larger production facilities. The increase in investment costs will add to the CO_2 mitigation costs.

Previous studies have discussed the potential role of CO_2 utilization to mitigate greenhouse gas emissions based on the amount of CO_2 utilized (Mac Dowell et al., 2017). The High Level Group of Scientific Advisors to the European Commission, for example, recently reported an estimated long-term utilization potential of 1-2 Gt per year (High Level Group of Scientific Advisors to the European Commission, 2018). Alternatively, the amount of CO_2 stored in chemicals has been used as a simplified indicator for the potential of CO_2 utilization to reduce greenhouse gas emissions (Mac Dowell et al., 2017). In contrast, the modeling approach proposed in this chapter provides a detailed assessment of greenhouse gas emission reductions due to disruptive changes through large-scale implementation of CO_2 utilization. The lack of models for assessing disruptive changes has recently been identified as a priority research need for life cycle assessment (Academies, 2019). The results show that the potential of a CO_2-based chemical industry to reduce greenhouse gas emissions is neither determined by the amount of CO_2 used, nor stored in chemicals, but by the substitution of conventional technologies. Therefore, estimates based on the amount of carbon used or stored should be interpreted with caution. Greenhouse gas emission reductions due to substitution have been demonstrated for individual CO_2-based technologies (Artz et al., 2018), but not for a large-scale CO_2-based global chemical industry. Finally, the analysis highlights the need to determine not only emission reductions compared with conventional technologies, but also to compare measures to reduce greenhouse gas emissions regarding their power-to-X efficiency.

D.5 Power-to-X efficiencies of e-mobility and power-to-Heat

The reductions of greenhouse gas emissions for the benchmark power-to-X technologies are determined based on power-to-X efficiencies by Sternberg and Bardow (2015). The power-to-X efficiencies are defined as greenhouse gas emissions reductions per megawatt-hour of electricity used and represent the slopes of the gray lines in Figure D.4. In Sternberg and Bardow (2015), the power-to-X efficiencies $GW_{redution}$ are determined based on the following equation:

$$GW_{redution} = M_{product} GW_{conv} - GW_{ESS} \tag{D.1}$$

The power-to-X technology uses 1 MWh of electricity to produce the amount $M_{product}$ of its product. Thereby, the conventional production of this amount of product is avoided, reducing greenhouse gas emissions by $M_{product} GW_{conv}$, where GW_{conv} represents the greenhouse gas emissions of conventional production of one product unit. GW_{ESS} represents the greenhouse gas emissions of the power-to-X technology using 1 MWh of electricity. For both power-to-X and substituted technologies, emissions due to operation and construction are considered.

Appendix E

Using the bottom-up model for oxygenated chemical production

Dedicated research on new processes to manufacture chemicals is needed to reduce greenhouse gas emissions. One particular field that is gaining increased attention from science and industry is the electrochemical production of not only hydrogen but also hydrocarbon products and oxygenates (de Luna et al., 2019). Currently, the anodic hydrocarbon-to-oxygenate research focuses on chemical products like aldehydes (Zheng et al., 2017; Chen et al., 2019), nitriles (Huang et al., 2018; Ding et al., 2020), tetrahydroisoquinolines (THIQs), or dihydroisoquinolines (DHIQs) (Huang et al., 2019).

Even though the production of these chemicals achieved Faradaic efficiencies above 90%, they are not the primary oxygenates that cause greenhouse gas emissions of the chemical industry (Ausfelder et al., 2013). Thus, in the field of oxygenated products via electrochemical routes, there is a mismatch between the chemical products actually being responsible for larger shares of greenhouse gas emissions and the focus of electrochemical research. In this appendix, electrochemical routes are assessed with regard to their potential reductions of greenhouse gas emissions. The focus is, however, on electrochemical routes that produce those chemicals having a large impact on global greenhouse gas emissions. By this means, the respective results will provide a guiding analysis for future research to produce oxygenated chemicals via electrochemical routes.

This chapter is currently under submission and has not yet been published.

E.1 Scope of the analysis in relation to the general bottom-up model

In this appendix, the bottom-up model is used to calculate the global greenhouse gas emissions of ammonia and oxygenate production. For this purpose, the same computational structure and bottom-up model are used for the plastics assessment (cf. Chapter 5).

The major difference is that for the analysis the production volumes for 18 chemicals and plastics are used (Table D.1 of Appendix D). In particular, Figure E.1 uses all production volumes for 18 chemicals and plastics (Table D.1), while Figures E.4 and E.5 use only the production volumes of ammonia, propylene oxide, phenol, acetone, ethylene glycol, terephthalic acid, and methanol. These chemicals represent the major oxygenates that lead to a large contribution to global greenhouse gas emissions of the chemical industry.

In addition to the existing bottom-up model, electrochemical oxygenate production technologies are included inside the bottom-up model based on ideal thermodynamic relations. Here the same procedure as described in Section 5.4 is used, except that the product yield is assumed 100% and $Q_{\mathrm{H,min}}$ is assumed to be supplied by electricity with 100% efficiency.

E.2 Greenhouse gas emissions of NH3 and oxygenates

This section first evaluates the global greenhouse gas emissions of 20 large-volume chemicals. These chemicals are responsible for >75% of global greenhouse gas emissions of the chemical industry (cf. Appendix D). They include ammonia, aliphatic hydrocarbons (ethylene, propylene), aromatic hydrocarbons (benzene, styrene, cumene, toluene, p-xylene and mixed xylenes), oxygenates (methanol, ethylene glycol, ethylene oxide, propylene oxide, terephthalic acid, phenol), polyolefins (polyethylene, polypropylene) as well as caprolactam, acrylonitrile, and vinyl chloride.

The model shows that the total cradle-to-gate greenhouse gas emissions of the global chemical industry will reach 2 Gt CO_2-eq per annum by 2030 (cf. Figure E.1). Over half of these emissions arise from the sum of NH_3 manufacture (23%) plus the partial oxidation of hydrocarbons to oxygenates (30%). These processes should therefore receive directed efforts from the scientific community to develop alternative production processes with reduced greenhouse gas emissions. The remainder arises from the production of polyolefins (26%), aromatics (11%), chlorine (3%), vinyl chloride (3%), and acrylonitrile (2%).

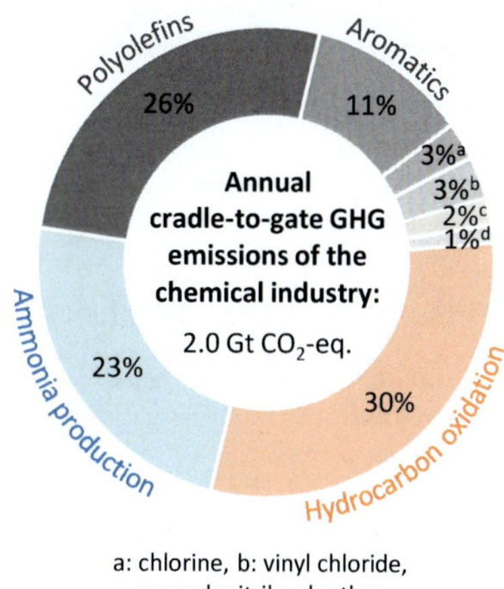

Figure E.1: Annual cradle-to-gate greenhouse gas emissions of the chemical industry in 2030.

A further breakdown of the annual cradle-to-gate emissions shows the major emission contributors (Figure E.2). For NH_3 manufacture, the bulk of associated greenhouse gas emissions is direct emissions from the steam reforming of methane with subsequent water-gas-shift reactions to produce the H_2 feedstock, which is then converted with nitrogen (N_2) to NH_3 via the Haber-Bosch reaction. This accounts for 0.4 Gt CO_2-eq. or 20% of the global chemical industry emissions (Figure E.1).

The direct emissions of hydrocarbon oxidation processes result from limited selectivities towards the target oxygenate products. This accounts for most of the 0.24 Gt CO_2-eq. or 12% of the global chemical industry emissions (Figure E.2), and arises primarily from methanol, ethylene glycol, and terephthalic acid production. These direct emissions arise from the oxidation of hydrocarbons all the way to CO_2; the results show that the development of alternative anodic hydrocarbon-to-oxygenate conversion methods with near-unity selectivities would reduce up to 0.11 Gt CO_2-eq. or 5.7%. The remaining 6.3% of direct emissions arise from the production of hydrocarbon feedstock via processes such as the steam cracking of naphtha and can also be reduced by lowering the consumption of hydrocarbon feedstock per ton of oxygenate.

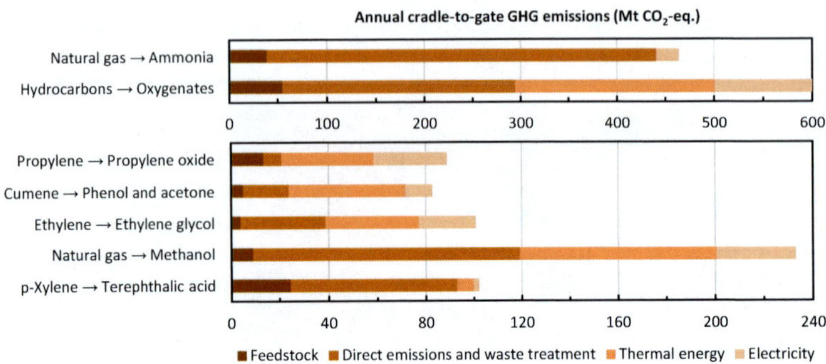

Figure E.2: Breakdown of annual cradle-to-gate emissions of NH_3 manufacture and hydrocarbon oxidations into feedstock, direct emissions and waste treatment, thermal energy, and electricity.

This maximizes the value of fossil-derived hydrocarbons and reduces feedstock-related emissions.

To reach the required temperatures and pressures for thermocatalytic reactions, feedstock needs to be heated and compressed - the consumed heat and electrical energy originates from fossil fuels and accounts for 0.21 and 0.12 Gt CO_2-eq. (i.e., 10% and 6%) of global greenhouse gas emissions associated with the chemical industry (Figure E.2). Although heat can be supplied by renewable-electricity-powered resistive heating, i.e., by retrofitting existing heating devices with electrode boilers, such electrical power-to-heat processes require vast amounts of electricity. The development of cathodic hydrogen evolution reactions and anodic hydrocarbon oxidations that can be conducted at ambient temperatures and pressures will reduce the greenhouse gas emissions associated with compression and heating.

E.3 Reduction potential of hydrocarbon-to-oxygenate conversions

As the bulk of greenhouse gas emissions associated with NH_3 manufacture arises from the production of H_2 feedstock, these emissions can be eliminated by using H_2 produced from renewable-electricity powered water electrolysis (Bhandari et al., 2014). However, the greenhouse gas reduction efficiency of today's water electrolysis

(WE) replacing steam-methane reforming is 0.10 t CO_2-eq per MWh of electricity (Figure E.3), which is lower than that of electric vehicles (EV) and heat pumps (HP) (Sternberg and Bardow, 2015). Thus, the limited renewable electricity resource should be allocated to water electrolysis only if such more efficient options are not available.

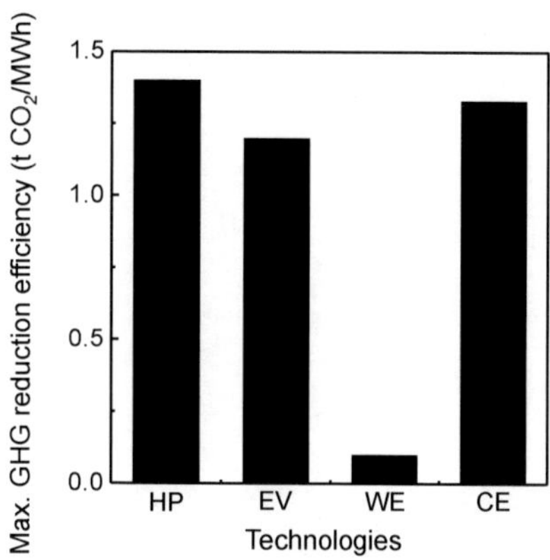

Figure E.3: Maximum greenhouse gas reduction efficiencies per MWh electricity of heat pumps (HP), electric vehicles (EV), water electrolysis (WE) and coupled electrolyzer (CE) technologies.

Renewable-energy-powered electrolyzer systems that couple cathodic hydrogen evolution reaction with anodic hydrocarbon-to-oxygenate conversion under ambient conditions can increase the greenhouse gas reduction efficiency up to 1.33 t CO_2-eq per MWh of electricity under ideal thermodynamic conditions (cf. Figure E.3), which is comparable to that of heat pumps and electric vehicles.

To assess the maximal reduction potential of the coupled electrochemical systems, the greenhouse gas emissions of the chemical industry are calculated for four scenarios: (1) the fossil-based NH_3 manufacture and partial oxidations of hydrocarbons to oxygenates (status quo), (2) employing renewable electricity without further changes to the manufacturing processes, (3) funneling all H_2 production to renewable-energy-powered water electrolysis and (4) the inclusion of renewable-energy-powered coupled cathodic hydrogen evolution reaction with anodic hydrocarbon-to-oxygenate conversion.

With Scenario 1 and approx. 1.1 Gt CO_2-eq emissions as the benchmark, Scenario 2 can reduce greenhouse gas emissions by 12%, while Scenario 3 can enable further reductions up to 21% (cf. Figure E.4). Scenario 4 can reduce greenhouse gas emissions even further by up to 47%. The remaining 53% of emissions are due to the operation of steam crackers and refineries to provide hydrocarbon feedstock and can be eliminated by replacing petrochemicals with cleaner and more sustainable feedstocks. This means that the coupled electrolyzer systems can avoid lock-in fossil technology by allowing for the transition from current fossil feedstocks to a completely net-zero situation with renewable carbon feedstock.

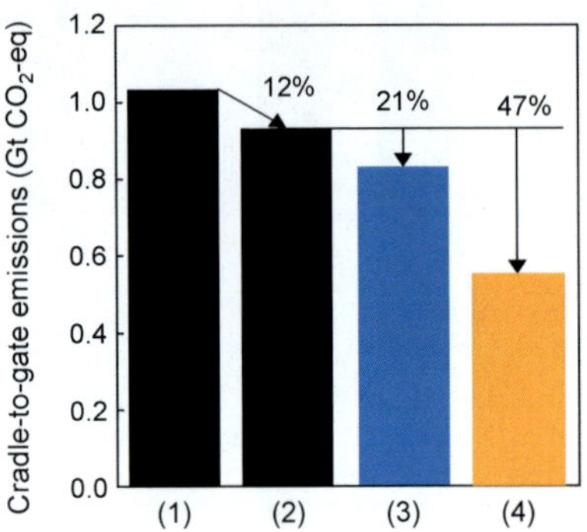

Figure E.4: Annual cradle-to-gate emissions of NH_3 manufacture and hydrocarbon-to-oxygenate conversions in four scenarios: (1) continuing to use fossil-based technologies, (2) employing renewable energy without further changes to the manufacturing processes, (3) funneling all H_2 production to renewable-energy-powered water electrolysis, and (4) the inclusion of renewable-energy-powered coupled electrolyzer technologies.

In addition, the greenhouse gas reduction potential of each scenario with electricity supply of different carbon intensities is calculated (cf. Figure E.5). The optimization approach minimizes the greenhouse gas emissions for the entire supply chain of the chemical industry. The results suggest that Scenario 3 can only reduce greenhouse gas emissions with an electricity supply of very low carbon intensity (i.e., <45 g CO_2-eq/kWh), which is not available in most industrialized countries today or even in 2030. For Scenario 4, the greenhouse gas emissions decline non-linearly; below 45 g CO_2-eq/kWh, the supply chain of NH_3 manufacture switches fully from fossil- to renewable-electricity-based hydrogen production. The optimization approach further determines that coupled electrolyzer technologies have the potential to reduce greenhouse gas emissions irrespective of the carbon intensity of the electricity supply. Greenhouse gas emissions can be reduced by up to 31% even with today's electricity supply in the United States or China (cf. Figure E.5). This clearly highlights the greenhouse gas reduction potential and the need for further research on anodic hydrocarbon-to-oxygenate conversions that can be coupled to hydrogen evolution reactions.

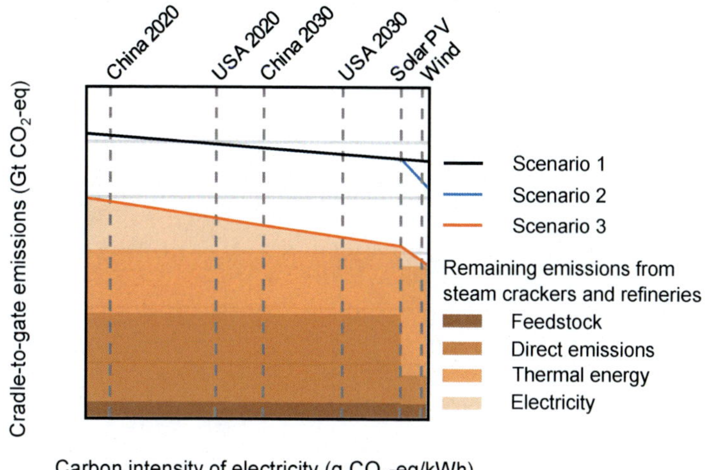

Figure E.5: Annual cradle-to-gate emissions of all scenarios with electricity supply of different carbon intensities

An analysis of state-of-art databases shows that 18% of global chemical sites currently conduct both NH_3 manufacture and partial oxidations of hydrocarbons to oxygenates (CarbonMinds, 2020). These sites can be targeted for the implementation of the coupled electrolyzer technologies in the short term. These co-production sites tend to be large production sites such as the one located in Ludwigshafen, Germany. In this "Verbund"-site, almost 1 Mt of ammonia and oxygenates are produced jointly and thus offer excellent conditions for the implementation of the coupled electrolyzer technologies.

This appendix showed that the use of anodic oxidations as a synthetic toolkit for hydrocarbon-to-oxygenate conversions is a burgeoning field with great potential to reduce greenhouse gas emissions of the chemical industry. Thus, dedicated efforts from the scientific community are needed to address the associated technical challenges in order to drive the electrification of the chemicals industry. By enabling the implementation of electrochemical oxidation reactions, up to 47% of the associated greenhouse gas emissions of ammonia and hydrocarbon oxygenates can be reduced if sufficient renewable electricity is available.

Thus, this appendix highlights the flexibility of the bottom-up model to not only assess mature technologies for plastic production but also provide estimations of greenhouse gas emission reductions by novel, early-development stage technologies like electrochemical processes.

Appendix F

Publications and student theses

F.1 List of journal publications

Artz, J., Müller, T. E., Thenert, K., Kleinekorte, J., Meys, R., Sternberg, A., Bardow, A., and Leitner, W. (2018). Sustainable conversion of carbon dioxide: An integrated review of catalysis and life cycle assessment. *Chemical reviews*, 118(2):434-504.

Meys, R., Kätelhön, and Bardow, A. (2019). Towards sustainable elastomers from CO2: life cycle assessment of carbon capture and utilization for rubbers. *Green Chemistry*, 21(12):3334-3342.

Kätelhön, A., Meys, R., Deutz, S., Suh, S. and Bardow, A. (2019). Climate change mitigation potential of carbon capture and utilization in the chemical industry. *Proceedings of the National Academy of Sciences of the United States of America*, 116(23):11187-11194.

Meys, R., Frick, F., Westhues, S., Sternberg, A., Klankermayer, J., and Bardow, A. (2020). Towards a circular economy for plastic packaging wastes - the environmental potential of chemical recycling. *Resources, Conservation and Recycling*, 162:105010.

Roh, K., Bardow, A., Bongartz, D., Burre, J., Chung, W., Deutz, S., Han, D., Heßelmann, M., Kohlhaas, Y., König, A., Lee, J. S., Meys, R., Völker, S., Wessling, M., Lee, J. H., and Mitsos, A. (2020). Early-stage evaluation of emerging CO2 utilization technologies at low technology readiness levels. *Green Chemistry*, 22(12):3842-3859.

Meys, R., Kätelhön, A., Bachmann, M., Winter, B., Zibunas, C., Suh, S. and Bardow, A. (2021). Achieving net-zero greenhouse gas emission plastics by a circular carbon economy. *Science*, 374(6563):71-76.

Bachmann, M., Kätelhön, A., Winter, B., Meys, R., Müller, L., and Bardow, A. (2021). Renewable Carbon Feedstock for Polymers - Environmental Benefits from Synergistic Use of Biomass and CO2. *Faraday discussions*.

Winter, B., Meys, R., and Bardow, A. (2021). Towards aromatics from biomass: Prospective life cycle assessment of bio-based aniline. *Journal of Cleaner Production*, 290:125818.

F.2 Student theses supervised during this work

Dörpinghaus, L. Prozesssimulation der Polyolherstellung aus Epoxiden und Kohlenstoffdioxid (2016). *In German*, Bachelor thesis. RWTH Aachen University.

Frick, F. Ökologische Bewertung von Kunststoffrecycling (2017). *In German*, Bachelor thesis. RWTH Aachen University.

Winter, B. Economic Optimization of Polyester Production from renewable Resources (2017). Master thesis. RWTH Aachen University.

Neumaier, L. Calculation of CO2 Emission Prices for Technology Changes Within the Chemical Industry Based on Optimization Models (2018). Bachelor thesis. RWTH Aachen University.

Schäfer, L. Ökologische Bewertung der Entsorgung von Abfällen der chemischen Industrie (2018). *In German*, Bachelor thesis. RWTH Aachen University.

Frankemölle, J. Fundierte ökologische Entscheidungen für die Chemieindustrie: Der Einfluss von LCA Methodik auf Technologiewahlen (2018). *In German*, Master thesis. RWTH Aachen University.

Romberg, H. (2019). Ökologische Bewertung von Elektrifizierungstechnologien zur Bereitstellung industrieller Prozesswärme mithilfe linearer Optimierungsmodelle *In German*, Bachelor thesis. RWTH Aachen University.

Dibos, S. and Pütz, O. Globale Umweltauswirkungen durch die Verwendung der besten verfügbaren Technologien zur Chemikalienproduktion (2019). *In German*, Project thesis. RWTH Aachen University.

Esser, M. Ecological Evaluation of Plastic Recycling in the Chemical Industry based on Life Cycle Optimization (2019). Master thesis. RWTH Aachen University.

Stellner, L. Environmental assessment of a fossil-free fuel and chemical industry based on linear optimization models (2019). Master thesis. RWTH Aachen University.

Dinges, V. and Ellinger, M. Application and Assessment of Simplified Process Design Methods for Generating Life Cycle Assessment Databases for Chemicals (2020). Project thesis. RWTH Aachen University.

Bibliography

Abella, J. P. and Bergerson, J. A. (2012). Model to investigate energy and greenhouse gas emissions implications of refining petroleum: impacts of crude quality and refinery configuration. *Environmental science & technology*, 46(24):13037–13047.

Academies, N. (2019). *Gaseous carbon waste streams utilization: Status and research needs*. A consensus study report of the National Academies of Sciences, Engineering, Medicine. The National Academies Press, Washington, DC.

Afeefy, H. Y., Liebman, J. F., and Stein, S. E. (2018). in NIST Chemistry WebBook.

Agora (2018). The future cost of electricity-based synthetic fuels.

Al-Kalbani, H., Xuan, J., García, S., and Wang, H. (2016). Comparative energetic assessment of methanol production from CO2: Chemical versus electrochemical process. *Applied Energy*, 165:1–13.

Al-Salem, S. M., Lettieri, P., and Baeyens, J. (2009). Recycling and recovery routes of plastic solid waste (PSW): a review. *Waste management (New York, N.Y.)*, 29(10):2625–2643.

Alagi, P., Ghorpade, R., Choi, Y. J., Patil, U., Kim, I., Baik, J. H., and Hong, S. C. (2017). Carbon dioxide-based polyols as sustainable feedstock of thermoplastic polyurethane for corrosion-resistant metal coating. *ACS Sustainable Chemistry & Engineering*, 5(5):3871–3881.

Alauddin, M., Choudhury, I. A., El Baradie, M. A., and Hashmi, M. (1995). Plastics and their machining: A review. *Journal of Materials Processing Technology*, 54(1-4):40–46.

Althaus, H.-J., Chudacoff, M., Hischier, R., Jungbluth, N., Osses, M., and Primas, A. (2007). *Life Cycle Inventories of Chemicals: ecoinvent report No. 8, V2.0*. EMPA Dübendorf, Swiss Centre for Life Cycle Inventories, Dübendorf.

Andrady, A. L. and Neal, M. A. (2009). Applications and societal benefits of plastics. *Philosophical transactions of the Royal Society of London. Series B, Biological sciences*, 364(1526):1977–1984.

Antelava, A., Damilos, S., Hafeez, S., Manos, G., Al-Salem, S. M., Sharma, B. K., Kohli, K., and Constantinou, A. (2019). Plastic Solid Waste (PSW) in the Context of Life Cycle Assessment (LCA) and Sustainable Management. *Environmental management*, 64(2):230–244.

Anuar Sharuddin, S. D., Abnisa, F., Daud, W. M. A. W., and Aroua, M. K. (2018). Pyrolysis of plastic waste for liquid fuel production as prospective energy resource. *IOP Conference Series: Materials Science and Engineering*, 334:012001.

Anuar Sharuddin, S. D., Abnisa, F., Wan Daud, W. M. A., and Aroua, M. K. (2016). A review on pyrolysis of plastic wastes. *Energy Conversion and Management*, 115:308–326.

Aresta, M., Dibenedetto, A., and Angelini, A. (2014). Catalysis for the valorization of exhaust carbon: from CO2 to chemicals, materials, and fuels. Technological use of CO2. *Chemical reviews*, 114(3):1709–1742.

Artz, J., Müller, T. E., Thenert, K., Kleinekorte, J., Meys, R., Sternberg, A., Bardow, A., and Leitner, W. (2018). Sustainable conversion of carbon dioxide: An integrated review of catalysis and life cycle assessment. *Chemical reviews*, 118(2):434–504.

Arvidsson, M., Morandin, M., and Harvey, S. (2014). Biomass gasification-based syngas production for a conventional oxo synthesis plant - process modeling, integration opportunities, and thermodynamic performance. *Enery & Fuels*, 28(6):4075–4087.

Arvidsson, M., Heyne, S., Morandin, M., and Harvey, S. (2012). Integration opportunities for substitute natural gas (SNG) production in an industrial process plant. *Chemical Engineering Transactions*, 29:331–336.

Asinger, F. (1986). *Methanol - Chemie- und Energierohstoff: Die Mobilisation der Kohle [In German]*. Springer Berlin Heidelberg, Berlin, Heidelberg.

Ausfelder, F., Bazzanella, A. M., van Brackle, H., Wilde, R., Beckmann, C., Mills, R., Rightor, E., Tam, C., Trudeau, N., and Botschek, P. (2013). *Technology Roadmap: Energy and GHG reductions in the chemical industry via catalytic processes*. IEA Publications International Energy Agency, Paris.

Bachmann, M., Kätelhön, A., Winter, B., Meys, R., Müller, L., and Bardow, A. (2021). Renewable Carbon Feedstock for Polymers - Environmental Benefits from Synergistic Use of Biomass and CO2. *Faraday discussions*.

Baena-Moreno, F. M., Rodríguez-Galán, M., Vega, F., Alonso-Fariñas, B., Vilches Arenas, L. F., and Navarrete, B. (2019). Carbon capture and utilization technologies: a literature review and recent advances. *Energy Sources, Part A: Recovery, Utilization, and Environmental Effects*, 41(12):1403 1433.

Barbarias, I., Lopez, G., Artetxe, M., Arregi, A., Bilbao, J., and Olazar, M. (2018). Valorisation of different waste plastics by pyrolysis and in-line catalytic steam reforming for hydrogen production. *Energy Conversion and Management*, 156:575–584.

Barnes, D. K. A., Galgani, F., Thompson, R. C., and Barlaz, M. (2009). Accumulation and fragmentation of plastic debris in global environments. *Philosophical transactions of the Royal Society of London. Series B, Biological sciences*, 364(1526):1985–1998.

Baumann, H. and Tillman, A.-M. (2004). *The hitch hikers's guide to LCA: An orientation in life cycle assessment methodology and application.* Studentlitteratur, Lund.

Bazzanella, A. M. and Ausfelder, F. (2017). *Low carbon energy and feedstock for the European chemical industry.* DECHEMA Gesellschaft für Chemische Technik und Biotechnologie e.V.

Berghout, N., Meerman, H., van den Broek, M., and Faaij, A. (2019). Assessing deployment pathways for greenhouse gas emissions reductions in an industrial plant - a case study for a complex oil refinery. *Applied Energy*, 236:354–378.

Bernardo, C. A., Simões, C. L., and Pinto, L. M. C. (2016). Environmental and economic life cycle analysis of plastic waste management options. A review. page 140001.

Bhandari, R., Trudewind, C. A., and Zapp, P. (2014). Life cycle assessment of hydrogen production via electrolysis - A review. *Journal of Cleaner Production*, 85:151–163.

BIO Intelligence Service (2013). Study on an increased mechanical recycling target for plastics: Final report prepared for plastics recyclers europe.

Black Rock (2021). Larry Fink's 2021 letter to CEOs, 08.03.2021. `https://www.blackrock.com/corporate/investor-relations/larry-fink-ceo-letter`. last accessed 08.03.2021.

Blanco, I., Ingrao, C., and Siracusa, V. (2020). Life-cycle assessment in the polymeric sector: A comprehensive review of application experiences on the italian scale. *Polymers*, 12(6).

Borrelle, S. B., Ringma, J., Law, K. L., Monnahan, C. C., Lebreton, L., McGivern, A., Murphy, E., Jambeck, J., Leonard, G. H., Hilleary, M. A., Eriksen, M., Possingham, H. P., de Frond, H., Gerber, L. R., Polidoro, B., Tahir, A., Bernard, M., Mallos, N., Barnes, M., and Rochman, C. M. (2020). Predicted growth in plastic waste exceeds efforts to mitigate plastic pollution. *Science (New York, N.Y.)*, 369(6510):1515–1518.

Bozell, J. J. and Petersen, G. R. (2010). Technology development for the production of biobased products from biorefinery carbohydrates - the us department of energy's "top 10" revisited. *Green Chemistry*, 12(4):539.

Brandt, R., Jaroma-Weiland, G., Neuer, G., Pohlmann, P., and Schreiber, E. (1998). Thermisches Verhalten von C/C-SiC [In German].

Brooks, A. L., Wang, S., and Jambeck, J. R. (2018). The chinese import ban and its impact on global plastic waste trade. *Science advances*, 4(6):eaat0131.

Brown, H. L. (1996). *Energy analysis of 108 industrial processes*. Fairmont Press, Lilburn, Ga.

Buckley, T. J. (1991). Calculation of higher heating values of biomass materials and waste components from elemental analyses. *Resources, Conservation and Recycling*, 5(4):329–341.

Burgess, D. R. (2018). in NIST Chemistry WebBook.

CarbonMinds (2020). Confidential notification based on internal cm.chemicals database.

Carta, D., Cao, G., and D'Angeli, C. (2003). Chemical recycling of poly(ethylene terephthalate) (PET) by hydrolysis and glycolysis. *Environmental science and pollution research international*, 10(6):390–394.

Cédric Philibert (2017a). Producing ammonia and fertilizers: New opportunities from renewables. iea.

Cédric Philibert (2017b). Renewable energy for industry: From green energy to green materials and fuels. iea.

Centi, G., Quadrelli, E. A., and Perathoner, S. (2013). Catalysis for CO2 conversion: a key technology for rapid introduction of renewable energy in the value chain of chemical industries. *Energy & Environmental Science*, 6(6):1711.

Cesaro, Z., Ives, M., Nayak-Luke, R., Mason, M., and Bañares-Alcántara, R. (2021). Ammonia to power: Forecasting the levelized cost of electricity from green ammonia in large-scale power plants. *Applied Energy*, 282:116009.

Cespi, D., Passarini, F., Vassura, I., and Cavani, F. (2016). Butadiene from biomass, a life cycle perspective to address sustainability in the chemical industry. *Green Chemistry*, 18(6):1625–1638.

Chamas, A., Moon, H., Zheng, J., Qiu, Y., Tabassum, T., Jang, J. H., Abu-Omar, M., Scott, S. L., and Suh, S. (2020). Degradation rates of plastics in the environment. *ACS Sustainable Chemistry & Engineering*, 8(9):3494–3511.

Chemical park operator (2018). Confidential information.

ChemSystems (2009). *PERP Report Abstract PERP07/08-6: Propylene Oxide.*

Chen, X., Zhong, X., Yuan, B., Li, S., Gu, Y., Zhang, Q., Zhuang, G., Li, X., Deng, S., and Wang, J.-g. (2019). Defect engineering of nickel hydroxide nanosheets by ostwald ripening for enhanced selective electrocatalytic alcohol oxidation. *Green Chemistry*, 21(3):578–588.

Clark, J. H., Farmer, T. J., Herrero-Davila, L., and Sherwood, J. (2016). Circular economy design considerations for research and process development in the chemical sciences. *Green Chemistry*, 18(14):3914–3934.

Concawe (January 2017). *Estimating the marginal CO2 intensities of EU refinery products*, volume no. 17, 1 of *Report / CONCAWE*. CONCAWE, Brussels.

Connolly, D., Lund, H., Mathiesen, B. V., Werner, S., Möller, B., Persson, U., Boermans, T., Trier, D., Østergaard, P. A., and Nielsen, S. (2014). Heat roadmap europe: Combining district heating with heat savings to decarbonise the eu energy system. *Energy Policy*, 65:475–489.

Consultic (2015). Produktion, Verarbeitung und Verwertung von Kunststoffen in Deutschland 2015 - Kurzfassung [In German].

Corsten, M., Worrell, E., Rouw, M., and van Duin, A. (2013). The potential contribution of sustainable waste management to energy use and greenhouse gas emission reduction in the netherlands. *Resources, Conservation and Recycling*, 77:13–21.

Creutzig, F., Ravindranath, N. H., Berndes, G., Bolwig, S., Bright, R., Cherubini, F., Chum, H., Corbera, E., Delucchi, M., Faaij, A., Fargione, J., Haberl, H., Heath, G., Lucon, O., Plevin, R., Popp, A., Robledo-Abad, C., Rose, S., Smith, P., Stromman, A., Suh, S., and Masera, O. (2015). Bioenergy and climate change mitigation: an assessment. *GCB Bioenergy*, 7(5):916–944.

Cruz, T. T. d., Perrella Balestieri, J. A., de Toledo Silva, J. M., Vilanova, M. R., Oliveira, O. J., and Ávila, I. (2021). Life cycle assessment of carbon capture and storage/utilization: From current state to future research directions and opportunities. *International Journal of Greenhouse Gas Control*, 108:103309.

Cuéllar-Franca, R. M. and Azapagic, A. (2015). Carbon capture, storage and utilisation technologies: A critical analysis and comparison of their life cycle environmental impacts. *Journal of CO2 Utilization*, 9:82–102.

Davidson, M. G., Furlong, R. A., and McManus, M. C. (2021). Developments in the life cycle assessment of chemical recycling of plastic waste - A review. *Journal of Cleaner Production*, 293:126163.

de Jong, E., Stichnothe, H., Bell, G., and Jorgensen, H. (2020). *Bio-based chemicals*. IEA Bioenergy.

de Luna, P., Hahn, C., Higgins, D., Jaffer, S. A., Jaramillo, T. F., and Sargent, E. H. (2019). What would it take for renewably powered electrosynthesis to displace petrochemical processes? *Science (New York, N.Y.)*, 364(6438).

Dehoust, G. and Christiani, J. (2012). *Analyse und Fortentwicklung der Verwertungsquoten für Wertstoffe (In German). Sammel- und Verwertungsquoten für Verpackungen und stoffgleiche Nichtverpackungen als Lenkungsinstrument zur Ressourcenschonung.*

Dehoust, G., Möck, A., Merz, C., and Gebhardt, P. (2016). In German: Umweltpotenziale der getrennten Erfassung und des Recyclings von Wertstoffen im Dualen System. `https://www.gruener-punkt.de/fileadmin/Dateien/Downloads/PDFs/16-09-21_OEko-Institut_Abschlussbericht_LCA-DSD.PDF`. last accessed 25.01.2022.

Deutz, S. and Bardow, A. (2021). Life-cycle assessment of an industrial direct air capture process based on temperature-vacuum swing adsorption. *Nature Energy*, 6(2):203–213.

DG RTD (2018a). *Novel carbon capture and utilisation technologies*, volume 4/2018 of *Scientific Advice Mechanism (SAM), Group of Chief Scientific Advisors, Scientific Opinion*. Publications Office, Luxembourg.

DG RTD (2018b). *Pathways to sustainable industries: Energy efficiency and CO2 utilisation*. Research & innovation projects for policy. Publications Office, Luxembourg.

Ding, Y., Miao, B.-Q., Li, S.-N., Jiang, Y.-C., Liu, Y.-Y., Yao, H.-C., and Chen, Y. (2020). Benzylamine oxidation boosted electrochemical water-splitting: Hydrogen and benzonitrile co-production at ultra-thin Ni2P nanomeshes grown on nickel foam. *Applied Catalysis B: Environmental*, 268:118393.

Dinh, C.-T., Burdyny, T., Kibria, M. G., Seifitokaldani, A., Gabardo, C. M., García de Arquer, F. P., Kiani, A., Edwards, J. P., de Luna, P., Bushuyev, O. S., Zou, C., Quintero-Bermudez, R., Pang, Y., Sinton, D., and Sargent, E. H. (2018). CO2 electroreduction to ethylene via hydroxide-mediated copper catalysis at an abrupt interface. *Science (New York, N.Y.)*, 360(6390):783–787.

Doka, G. (2003). Life cycle inventories of waste treatment services: ecoinvent report no. 13.

Doka, G. (2013). Updates to life cycle inventories of waste treatment services - part ii: waste incineration.

Dong, Y., Zhao, Y., Hossain, M. U., He, Y., and Liu, P. (2021). Life cycle assessment of vehicle tires: A systematic review. *Cleaner Environmental Systems*, page 100033.

DSD (2018). Spezifikation der Wertstofffraktion (in German).

Eceiza, A., Martin, M. D., de La Caba, K., Kortaberria, G., Gabilondo, N., Corcuera, M. A., and Mondragon, I. (2008). Thermoplastic polyurethane elastomers based on polycarbonate diols with different soft segment molecular weight and chemical structure: Mechanical and thermal properties. *Polymer Engineering & Science*, 48(2):297–306.

Ecoinvent (2020). ecoinvent data v. 3.6.

Edenhofer, O., editor (2012). *Renewable energy sources and climate change mitigation: Special report of the Intergovernmental Panel on Climate Change*. Cambridge University Press, Cambridge.

Ekvall, T. and Tillman, A.-M. (1997). Open-loop recycling: Criteria for allocation procedures. *The International Journal of Life Cycle Assessment*, 2(3):155–162.

El-Houjeiri, H. M., Brandt, A. R., and Duffy, J. E. (2013). Open-source LCA tool for estimating greenhouse gas emissions from crude oil production using field characteristics. *Environmental science & technology*, 47(11):5998–6006.

Elordi, G., Olazar, M., Lopez, G., Amutio, M., Artetxe, M., Aguado, R., and Bilbao, J. (2009). Catalytic pyrolysis of hdpe in continuous mode over zeolite catalysts in a conical spouted bed reactor. *Journal of Analytical and Applied Pyrolysis*, 85(1-2):345–351.

Elvers, B. and Ullmann, F., editors (2011). *Ullmann's encyclopedia of industrial chemistry*. Wiley-VCH, Weinheim, 7. comp. rev. ed. edition.

Eriksson, O., Carlsson Reich, M., Frostell, B., Björklund, A., Assefa, G., Sundqvist, J.-O., Granath, J., Baky, A., and Thyselius, L. (2005). Municipal solid waste management from a systems perspective. *Journal of Cleaner Production*, 13(3):241–252.

Eriksson, O. and Finnveden, G. (2009). Plastic waste as a fuel - CO2-neutral or not? *Energy & Environmental Science*, 2(9):907.

Eriksson, O. and Finnveden, G. (2017). Energy Recovery from Waste Incineration - The Importance of Technology Data and System Boundaries on CO2 Emissions. *Energies*, 10(4):539.

Eriksson, O., Finnveden, G., Ekvall, T., and Björklund, A. (2007). Life cycle assessment of fuels for district heating: A comparison of waste incineration, biomass- and natural gas combustion. *Energy Policy*, 35(2):1346–1362.

Eş, I., Mousavi Khaneghah, A., Barba, F. J., Saraiva, J. A., Sant'Ana, A. S., and Hashemi, S. M. B. (2018). Recent advancements in lactic acid production - a review. *Food research international (Ottawa, Ont.)*, 107:763–770.

Escobar, N. and Laibach, N. (2021). Sustainability check for bio-based technologies: A review of process-based and life cycle approaches. *Renewable and Sustainable Energy Reviews*, 135:110213.

European Commission (2010). Guidance on Interpretation of Annex I of the EU ETS Directive (excl. aviation activities). https://ec.europa.eu/clima/system/files/2016-11/guidance_interpretation_en.pdf. last accessed 25.01.2022.

European Commission (2015). Closing the loop - An EU action plan for the Circular Economy.

European Commission. Joint Research Centre. (2007). *Best Available Techniques (BAT) reference document for the production of polymers.* Publications Office.

European Commission. Joint Research Centre. (2017). *Best Available Techniques (BAT) reference document for the production of large volume organic chemicals.* Publications Office.

Eurostat (2018). Renewable energy statistics. http://ec.europa.eu/eurostat/statistics-explained/index.php/Renewable_energy_statistics. last accessed June 2018.

Farla, J. C., Hendriks, C. A., and Blok, K. (1995). Carbon dioxide recovery from industrial processes. *Energy Conversion and Management*, 36(6-9):827–830.

Fasihi, M., Bogdanov, D., and Breyer, C. (2016). Techno-Economic Assessment of Power-to-Liquids (PtL) Fuels Production and Global Trading Based on Hybrid PV-Wind Power Plants. *Energy Procedia*, 99:243–268.

Finkbeiner, M. and Bach, V. (2021). Life cycle assessment of decarbonization options - towards scientifically robust carbon neutrality. *The International Journal of Life Cycle Assessment.*

Foundation, E. M. (2016). The new plastics economy: Rethinking the future of plastics.

Fragkos, P. (2011). Energy roadmap 2050 - impact assessment and scenario analysis.

Frischknecht, R., Althaus, H.-J., Bauer, C., Doka, G., Heck, T., Jungbluth, N., Kellenberger, D., and Nemecek, T. (2007). The environmental relevance of capital goods in life cycle assessments of products and services. *Int J LCA*, (13):7–17.

Fuss, S., Lamb, W. F., Callaghan, M. W., Hilaire, J., Creutzig, F., Amann, T., Beringer, T., de Oliveira Garcia, W., Hartmann, J., Khanna, T., Luderer, G., Nemet, G. F., Rogelj, J., Smith, P., Vicente, J. L. V., Wilcox, J., del Mar Zamora Dominguez, M., and Minx, J. C. (2018). Negative emissions - part 2: Costs, potentials and side effects. *Environmental Research Letters*, 13(6):063002.

Gabrielli, P., Gazzani, M., and Mazzotti, M. (2020). The Role of Carbon Capture and Utilization, Carbon Capture and Storage, and Biomass to Enable a Net-Zero-CO2 Emissions Chemical Industry. *Industrial & Engineering Chemistry Research*, 59(15):7033–7045.

Galán-Martín, Á., Tulus, V., Díaz, I., Pozo, C., Pérez-Ramírez, J., and Guillén-Gosálbez, G. (2021). Sustainability footprints of a renewable carbon transition for the petrochemical sector within planetary boundaries. *One Earth*, 4(4):565–583.

Gao, J., Jia, C., and Liu, B. (2017). Direct and selective hydrogenation of CO2 to ethylene and propene by bifunctional catalysts. *Catalysis Science & Technology*, 7(23):5602–5607.

Gao, W., Hundertmark, T., Simons, T., Wallach, J., and Witte, C. (2021). Plastics recycling: Using an economic-feasibility lens to select the next moves. `https://www.mckinsey.com/industries/chemicals/our-insights/plastics-recycling-using-an-economic-feasibility-lens-to-select-the-next-moves`. last accessed 11.01.2021.

Garcia-Garcia, G., Fernandez, M. C., Armstrong, K., Woolass, S., and Styring, P. (2021). Analytical review of life-cycle environmental impacts of carbon capture and utilization technologies. *ChemSusChem*, 14(4):995–1015.

Geisler, G., Hofstetter, T. B., and Hungerbühler, K. (2004). Production of fine and speciality chemicals: procedure for the estimation of lcis. *The International Journal of Life Cycle Assessment*, 9(2):101–113.

Gerssen-Gondelach, S. J., Saygin, D., Wicke, B., Patel, M. K., and Faaij, A. (2014). Competing uses of biomass: Assessment and comparison of the performance of bio-based heat, power, fuels and materials. *Renewable and Sustainable Energy Reviews*, 40:964–998.

Geyer, R., Jambeck, J. R., and Law, K. L. (2017). Production, use, and fate of all plastics ever made. *Science advances*, 3(7):e1700782.

Geyer, R., Kuczenski, B., Zink, T., and Henderson, A. (2016). Common misconceptions about recycling. *Journal of Industrial Ecology*, 20(5):1010–1017.

Glushko Thermocenter (2018). in NIST Chemistry WebBook.

Goedkoop, M., Heijungs, R., Huijbregts, M., Schryver, A. D., Struijs, J., and van Zelm, R. (2013). Recipe 2008: A life cycle impact assessment method which comprises harmonised category indicators at the midpoint and the end point level: First edition (version 1.08), report i: Characterisation.

Gollakota, A., Kishore, N., and Gu, S. (2018). A review on hydrothermal liquefaction of biomass. *Renewable and Sustainable Energy Reviews*, 81:1378–1392.

Gomes, T. S., Visconte, L. L. Y., and Pacheco, E. B. A. V. (2019). Life cycle assessment of polyethylene terephthalate packaging: An overview. *Journal of Polymers and the Environment*, 27(3):533–548.

Gordon, D., Brandt, A., Bergerson, J., and Koomey, J. (2015). *Creating a global oil-climate index.*

Guillén-Gosálbez, G. and Grossmann, I. (2010). A global optimization strategy for the environmentally conscious design of chemical supply chains under uncertainty in the damage assessment model. *Computers & Chemical Engineering*, 34(1):42–58.

Guo, X., Xin, J., Lu, X., Ren, B., and Zhang, S. (2015). Preparation of 1,4-cyclohexanedimethanol by selective hydrogenation of a waste pet monomer bis(2-hydroxyethylene terephthalate). *RSC Advances*, 5(1):485–492.

Haaf, M., Anantharaman, R., Roussanaly, S., Ströhle, J., and Epple, B. (2020). CO2 capture from waste-to-energy plants: Techno-economic assessment of novel integration concepts of calcium looping technology. *Resources, Conservation and Recycling*, 162:104973.

Hannula, I. and Kurkela, E. (2010). A semi-empirical model for pressurised air-blown fluidized-bed gasification of biomass. *Bioresource technology*, 101(12):4608–4615.

Hannula, I. and Kurkela, E. (2012). A parametric modelling study for pressurised steam/O2-blown fluidised-bed gasification of wood with catalytic reforming. *Biomass and Bioenergy*, 38:58–67.

Happ, M., Duffy, J., Wilson, G. J., Pask, S. D., Buding, H., and Ostrowicki, A. (2011). Rubber, 8. synthesis by polymer modification. In *Ullmann's Encyclopedia of Industrial Chemistry*. American Cancer Society.

Hatti-Kaul, R., Törnvall, U., Gustafsson, L., and Börjesson, P. (2007). Industrial biotechnology for the production of bio-based chemicals - a cradle-to-grave perspective. *Trends in biotechnology*, 25(3):119–124.

Hauschild, M. and Huijbregts, M. A. J., editors (2015). *Life cycle impact assessment.* LCA compendium, the complete world of life cycle assessment. Springer, Dordrecht.

Haward, M. (2018). Plastic pollution of the world's seas and oceans as a contemporary challenge in ocean governance. *Nature communications*, 9(1):667.

Heijungs, R. and Suh, S. (2002). *The Computational Structure of Life Cycle Assessment*, volume 11 of *Eco-Efficiency in Industry and Science*. Springer, Dordrecht.

Hermann, B. G., Blok, K., and Patel, M. K. (2007). Producing bio-based bulk chemicals using industrial biotechnology saves energy and combats climate change. *Environmental science & technology*, 41(22):7915–7921.

High Level Group of Scientific Advisors to the European Commission (2018). Novel carbon capture and utilization technologies.

Hischier, R., Hellweg, S., Capello, C., and Primas, A. (2005). Establishing life cycle inventories of chemicals based on differing data availability (9 pp). *The International Journal of Life Cycle Assessment*, 10(1):59–67.

Holladay, J. E., White, J. F., Bozell, J. J., and Johnson, D. (2007). Top value-added chemicals from biomass-volume ii - results of screening for potential candidates from biorefinery lignin.

Hong, M. and Chen, E. Y.-X. (2017). Chemically recyclable polymers: a circular economy approach to sustainability. *Green Chemistry*, 19(16):3692–3706.

Honus, S., Kumagai, S., Fedorko, G., Molnár, V., and Yoshioka, T. (2018a). Pyrolysis gases produced from individual and mixed PE, PP, PS, PVC, and PET - Part I: Production and physical properties. *Fuel*, 221:346–360.

Honus, S., Kumagai, S., Molnár, V., Fedorko, G., and Yoshioka, T. (2018b). Pyrolysis gases produced from individual and mixed PE, PP, PS, PVC, and PET - Part II: Fuel characteristics. *Fuel*, 221:361–373.

Hoppe, W., Thonemann, N., and Bringezu, S. (2018). Life cycle assessment of carbon dioxide-based production of methane and methanol and derived polymers. *Journal of Industrial Ecology*, 22(2):327–340.

HTP GmbH & Co. KG (2017). Confidential notification based on internal project reports.

Huang, C., Huang, Y., Liu, C., Yu, Y., and Zhang, B. (2019). Integrating Hydrogen Production with Aqueous Selective Semi-Dehydrogenation of Tetrahydroisoquinolines over a Ni2 P Bifunctional Electrode. *Angewandte Chemie (International ed. in English)*, 58(35):12014–12017.

Huang, Y., Chong, X., Liu, C., Liang, Y., and Zhang, B. (2018). Boosting Hydrogen Production by Anodic Oxidation of Primary Amines over a NiSe Nanorod Electrode. *Angewandte Chemie (International ed. in English)*, 57(40):13163–13166.

Huijbregts, M., Steinmann, Z., Elshout, P., Stam, G., Verones, F., Vieira, M., Hollander, A., Zijp, M., and van Zelm, R. (2016). Recipe 2016: A harmonized life cycle impact assessment method at midpoint and endpoint level: Report i: Characterization. www.rivm.nl.

Humbird, D., Davis, R., Tao, L., Kinchin, C., Hsu, D., Aden, A., Schoen, P., Lukas, J., Olthof, B., Worley, M., Sexton, D., and Dudgeon, D. (2011). Process design and economics for biochemical conversion of lignocellulosic biomass to ethanol.

ICIS (2018). Supply and demand database.

IEA, editor (2018a). *The Future of Petrochemicals*. IEA Publications International Energy Agency.

IEA (2018b). World energy outlook 2018. https://www.iea.org/weo/.

IEA (2020). Energy technology perspectives 2020. https://www.iea.org/reports/energy-technology-perspectives-2020. last accessed 25.10.2020.

IEA and CSI (2018). Technology roadmap low-carbon transition in the cement industry.

IHS Markit, editor (2018). *Process Economics Program (PEP) Yearbook*.

ILCD (2010a). *International Reference Life Cycle Data System (ILCD) Handbook - General guide for Life Cycle Assessment -Detailed guidance*, volume 24708 of *EUR. Scientific and technical research series*. Publications Office, Luxembourg, first ed. edition.

ILCD (2010b). *International Reference Life Cycle Data System (ILCD) Handbook - General guide for Life Cycle Assessment: Provisions and action steps*, volume 24378 of *EUR. Scientific and technical research series*. Publications Office, Luxembourg, first ed. edition.

International Renewable Energy Agency (2016). Boosting biofuels: Sustainable paths to greater energy security.

International Renewable Energy Agency (2019). Global energy transformation. a roadmap to 2050.

IPCC (2014). *Climate change 2013: The physical science basis : Working Group I contribution to the Fifth assessment report of the Intergovernmental Panel on Climate Change.* Cambridge University Press, Cambridge.

IPCC (2021). *Glossary.* IPCC.

Isaksson, J., Pettersson, K., Mahmoudkhani, M., Åsblad, A., and Berntsson, T. (2012). Integration of biomass gasification with a Scandinavian mechanical pulp and paper mill – Consequences for mass and energy balances and global CO2 emissions. *Energy*, 44(1):420–428.

Isensee, V. and Valdes, L. (2015). GSDR 2015 Brief - Marine Litter: Microplastics.

ISO 14040 (2021). Environmental management - life cycle assessment - principles and framework.

ISO 14044 (2021). Environmental management - life cycle assessment - requirements and guidlines.

Jambeck, J. R., Geyer, R., Wilcox, C., Siegler, T. R., Perryman, M., Andrady, A., Narayan, R., and Law, K. L. (2015). Marine pollution. plastic waste inputs from land into the ocean. *Science (New York, N.Y.)*, 347(6223):768–771.

Jeswani, H., Krüger, C., Russ, M., Horlacher, M., Antony, F., Hann, S., and Azapagic, A. (2021). Life cycle environmental impacts of chemical recycling via pyrolysis of mixed plastic waste in comparison with mechanical recycling and energy recovery. *Science of The Total Environment*, page 144483.

Jing, L., El-Houjeiri, H. M., Monfort, J.-C., Brandt, A. R., Masnadi, M. S., Gordon, D., and Bergerson, J. A. (2020). Carbon intensity of global crude oil refining and mitigation potential. *Nature Climate Change*, 10(6):526–532.

Johansson, D., Rootzén, J., Berntsson, T., and Johnsson, F. (2012). Assessment of strategies for CO2 abatement in the European petroleum refining industry. *Energy*, 42(1):375–386.

Jones, M. B. and Albanito, F. (2020). Can biomass supply meet the demands of bioenergy with carbon capture and storage (BECCS)? *Global change biology*.

Jung, J., von der Assen, N., and Bardow, A. (2013). Comparative LCA of multi-product processes with non-common products: a systematic approach applied to chlorine electrolysis technologies. *The International Journal of Life Cycle Assessment*, 18(4):828–839.

Kakadellis, S. and Harris, Z. M. (2020). Don't scrap the waste: The need for broader system boundaries in bioplastic food packaging life cycle assessment - a critical review. *Journal of Cleaner Production*, 274:122831.

Kalargaris, I., Tian, G., and Gu, S. (2017). The utilisation of oils produced from plastic waste at different pyrolysis temperatures in a di diesel engine. *Energy*, 131:179–185.

Kätelhön, A., Bardow, A., and Suh, S. (2016). Stochastic technology choice model for consequential life cycle assessment. *Environmental science & technology*, 50(23):12575–12583.

Kaza, S., Yao, L. C., Bhada-Tata, P., and van Woerden, F. (2018). *What a Waste 2.0: A Global Snapshot of Solid Waste Management to 2050*. Washington, DC: World Bank.

Kiss, A. A., Pragt, J. J., Vos, H. J., Bargeman, G., and de Groot, M. T. (2016). Novel efficient process for methanol synthesis by CO2 hydrogenation. *Chemical Engineering Journal*, 284:260–269.

Kondratenko, E. V., Mul, G., Baltrusaitis, J., Larrazábal, G. O., and Pérez-Ramírez, J. (2013). Status and perspectives of CO2 conversion into fuels and chemicals by catalytic, photocatalytic and electrocatalytic processes. *Energy & Environmental Science*, 6(11):3112.

Kriescher, S. M., Kugler, K., Hosseiny, S. S., Gendel, Y., and Wessling, M. (2015). A membrane electrode assembly for the electrochemical synthesis of hydrocarbons from CO2(g) and H2O(g). *Electrochemistry Communications*, 50:64–68.

Kümmerer, K., Clark, J. H., and Zuin, V. G. (2020). Rethinking chemistry for a circular economy. *Science (New York, N.Y.)*, 367(6476):369–370.

Langanke, J., Wolf, A., Hofmann, J., Böhm, K., Subhani, M. A., Müller, T. E., Leitner, W., and Gürtler, C. (2014). Carbon dioxide (CO2) as sustainable feedstock for polyurethane production. *Green Chem*, 16(4):1865–1870.

Lau, W. W. Y., Shiran, Y., Bailey, R. M., Cook, E., Stuchtey, M. R., Koskella, J., Velis, C. A., Godfrey, L., Boucher, J., Murphy, M. B., Thompson, R. C., Jankowska, E., Castillo Castillo, A., Pilditch, T. D., Dixon, B., Koerselman, L., Kosior, E., Favoino, E., Gutberlet, J., Baulch, S., Atreya, M. E., Fischer, D., He, K. K., Petit, M. M., Sumaila, U. R., Neil, E., Bernhofen, M. V., Lawrence, K., and Palardy, J. E. (2020). Evaluating scenarios toward zero plastic pollution. *Science (New York, N.Y.)*, 369(6510):1455–1461.

Laurent, A., Bakas, I., Clavreul, J., Bernstad, A., Niero, M., Gentil, E., Hauschild, M. Z., and Christensen, T. H. (2014a). Review of LCA studies of solid waste management systems-part I: lessons learned and perspectives. *Waste management (New York, N.Y.)*, 34(3):573–588.

Laurent, A., Clavreul, J., Bernstad, A., Bakas, I., Niero, M., Gentil, E., Christensen, T. H., and Hauschild, M. Z. (2014b). Review of LCA studies of solid waste management systems-part II: methodological guidance for a better practice. *Waste management (New York, N.Y.)*, 34(3):589–606.

Lazarevic, D., Aoustin, E., Buclet, N., and Brandt, N. (2010). Plastic waste management in the context of a european recycling society: Comparing results and uncertainties in a life cycle perspective. *Resources, Conservation and Recycling*, 55(2):246–259.

Leal Filho, W., Saari, U., Fedoruk, M., Iital, A., Moora, H., Klöga, M., and Voronova, V. (2019). An overview of the problems posed by plastic products and the role of extended producer responsibility in europe. *Journal of Cleaner Production*, 214:550–558.

Lebreton, L. and Andrady, A. (2019). Future scenarios of global plastic waste generation and disposal. *Palgrave Communications*, 5(1).

Levi, P. G. and Cullen, J. M. (2018). Mapping global flows of chemicals: From fossil fuel feedstocks to chemical products. *Environmental science & technology*, 52(4):1725–1734.

Li, A., Li, X., Li, S., Ren, Y., Shang, N., Chi, Y., Yan, J., and Cen, K. (1999). Experimental studies on municipal solid waste pyrolysis in a laboratory-scale rotary kiln. *Energy*, 24(3):209–218.

Linstrom, P. (2018). NIST Chemistry WebBook, NIST Standard Reference Database 69.

Lipski, A. and Hopmann, C. (2017). Optimisation of the Compound Quality of CO2-based Rubber Compounds. *KGK Rubberpoint.*

Liu, Z., Adams, M., Cote, R. P., Chen, Q., Wu, R., Wen, Z., Liu, W., and Dong, L. (2018). How does circular economy respond to greenhouse gas emissions reduction: An analysis of Chinese plastic recycling industries. *Renewable and Sustainable Energy Reviews*, 91:1162–1169.

Lopez, G., Artetxe, M., Amutio, M., Bilbao, J., and Olazar, M. (2017). Thermochemical routes for the valorization of waste polyolefinic plastics to produce fuels and chemicals. A review. *Renewable and Sustainable Energy Reviews*, 73:346–368.

Mac Dowell, N., Fennell, P. S., Shah, N., and Maitland, G. C. (2017). The role of CO2 capture and utilization in mitigating climate change. *Nature Climate Change*, 7(4):243–249.

Maga, D., Hiebel, M., and Thonemann, N. (2019). Life cycle assessment of recycling options for polylactic acid. *Resources, Conservation and Recycling*, 149:86–96.

Maranghi, S. and Brondi, C. (2020). *Life Cycle Assessment in the Chemical Product Chain: Challenges, Methodological Approaches and Applications*. Springer International Publishing and Imprint: Springer, Cham, 1st ed. 2020 edition.

Marson, A., Masiero, M., Modesti, M., Scipioni, A., and Manzardo, A. (2021). Life cycle assessment of polyurethane foams from polyols obtained through chemical recycling. *ACS omega*, 6(2):1718–1724.

Masnadi, M. S., El-Houjeiri, H. M., Schunack, D., Li, Y., Englander, J. G., Badahdah, A., Monfort, J.-C., Anderson, J. E., Wallington, T. J., Bergerson, J. A., Gordon, D., Koomey, J., Przesmitzki, S., Azevedo, I. L., Bi, X. T., Duffy, J. E., Heath, G. A., Keoleian, G. A., McGlade, C., Meehan, D. N., Yeh, S., You, F., Wang, M., and Brandt, A. R. (2018). Global carbon intensity of crude oil production. *Science (New York, N.Y.)*, 361(6405):851–853.

Matzen, M. and Demirel, Y. (2016). Methanol and dimethyl ether from renewable hydrogen and carbon dioxide: Alternative fuels production and life-cycle assessment. *Journal of Cleaner Production*, 139:1068–1077.

McCoy, M. (2020). New polyol plant upcycles pet bottles. *C&EN Global Enterprise*, 98(42):12.

Michaud, J.-C., Farrant, L., Jan, O., Kjær, B., and Bakas, I. (2010). *Environmental benefits of recycling - 2010 update.*

Milne, B. J., Behie, L. A., and Berruti, F. (1999). Recycling of waste plastics by ultrapyrolysis using an internally circulating fluidized bed reactor. *Journal of Analytical and Applied Pyrolysis*, 51(1-2):157–166.

Mobley, P. D., Peters, J. E., Akunuri, N., Hlebak, J., Gupta, V., Zheng, Q., Zhou, S. J., and Lail, M. (2017). Utilization of CO2 for Ethylene Oxide. *Energy Procedia*, 114:7154–7161.

Montazeri, M., Zaimes, G. G., Khanna, V., and Eckelman, M. J. (2016). Meta-analysis of life cycle energy and greenhouse gas emissions for priority biobased chemicals. *ACS Sustainable Chemistry & Engineering*, 4(12):6443–6454.

Müller, L. J., Kätelhön, A., Bachmann, M., Zimmermann, A., Sternberg, A., and Bardow, A. (2020a). A guideline for life cycle assessment of carbon capture and utilization. *Frontiers in Energy Research*, 8.

Müller, L. J., Kätelhön, A., Bringezu, S., McCoy, S., Suh, S., Edwards, R., Sick, V., Kaiser, S., Cuéllar-Franca, R., El Khamlichi, A., Lee, J. H., von der Assen, N., and Bardow, A. (2020b). The carbon footprint of the carbon feedstock CO2. *Energy & Environmental Science*, 13(9):2979–2992.

Muñoz, I., Flury, K., Jungbluth, N., Rigarlsford, G., i Canals, L. M., and King, H. (2014). Life cycle assessment of bio-based ethanol produced from different agricultural feedstocks. *The International Journal of Life Cycle Assessment*, 19(1):109–119.

Muthukumaran, P., Suresh Babu, P., Karthikeyan, S., Kamaraj, M., and Aravind, J. (2021). Tailored natural polymers: A useful eco-friendly sustainable tool for the mitigation of emerging pollutants: A review. *International Journal of Environmental Science and Technology.*

Nat Sustain (2018). Closing the plastics loop. *Nature Sustainability*, 1(5):205.

Navarro, J., Centeno, M., Laguna, O., and Odriozola, J. (2018). Policies and Motivations for the CO2 Valorization through the Sabatier Reaction Using Structured Catalysts. A Review of the Most Recent Advances. *Catalysts*, 8(12):578.

OECD (2018). Improving markets for recycled plastics: Trends, prospects and policy responses.

Olah, G. A., Goeppert, A., and Prakash, G. K. S. (2009). *Beyond oil and gas: The methanol economy*. Wiley-VCH, Weinheim, 2nd updated and enlarged ed. edition.

Opia, A. C., Hamid, M. K. B. A., Syahrullail, S., Rahim, A. B. A., and Johnson, C. A. (2021). Biomass as a potential source of sustainable fuel, chemical and tribological materials - Overview. *Materials Today: Proceedings*, 39:922–928.

Paszun, D. and Spychaj, T. (1997). Chemical recycling of poly(ethylene terephthalate). *Industrial & Engineering Chemistry Research*, 36(4):1373–1383.

Patel, A. D., Meesters, K., den Uil, H., de Jong, E., Blok, K., and Patel, M. K. (2012). Sustainability assessment of novel chemical processes at early stage: application to biobased processes. *Energy & Environmental Science*, 5(9):8430.

Patel, M., von Thienen, N., Jochem, E., and Worrell, E. (2000). Recycling of plastics in Germany. *Resources, Conservation and Recycling*, 29(1-2):65–90.

Pawelzik, P., Carus, M., Hotchkiss, J., Narayan, R., Selke, S., Wellisch, M., Weiss, M., Wicke, B., and Patel, M. K. (2013). Critical aspects in the life cycle assessment (LCA) of bio-based materials - Reviewing methodologies and deriving recommendations. *Resources, Conservation and Recycling*, 73:211–228.

Pelton, R. (2019). Spatial greenhouse gas emissions from US county corn production. *The International Journal of Life Cycle Assessment*, 24(1):12–25.

Pérez-Fortes, M. and Tzimas, E. (2016). *Techno-economic and environmental evaluation of CO2 utilisation for fuel production: Synthesis of methanol and formic acid*, volume 27629 of *EUR, Scientific and technical research series*. Publications Office, Luxembourg.

Pérez-Uresti, S. I., Martín, M., and Jiménez-Gutiérrez, A. (2019). Estimation of renewable-based steam costs. *Applied Energy*, 250:1120–1131.

Perugini, F., Mastellone, M. L., and Arena, U. (2005). A life cycle assessment of mechanical and feedstock recycling options for management of plastic packaging wastes. *Environmental Progress*, 24(2):137–154.

Phyllis2 (2019). database for biomass and waste. `https://phyllis.nl/`.

Picuno, C., Alassali, A., Chong, Z. K., and Kuchta, K. (2021). Flows of post-consumer plastic packaging in Germany: An MFA-aided case study. *Resources, Conservation and Recycling*, 169:105515.

PlasticsEurope, editor (2018). *Plastics - the Facts 2018. An analysis of European plastics production, demand and waste data.*

PlasticsInsights (2016). Polyamide production, pricing and market demand. `https://www.plasticsinsight.com/resin-intelligence/resin-prices/polyamide/production`. last accessed 22.02.2021.

Posen, I. D., Jaramillo, P., Landis, A. E., and Griffin, W. M. (2017). Greenhouse gas mitigation for U.S. plastics production: energy first, feedstocks later. *Environmental Research Letters*, 12(3):034024.

Prognos AG (2008). Resource savings and CO2 reduction potential in waste management in Europe and the possible contribution to the CO2 reduction target in 2020.

Pyl, S. P., Schietekat, C. M., Reyniers, M.-F., Abhari, R., Marin, G. B., and van Geem, K. M. (2011). Biomass to olefins: Cracking of renewable naphtha. *Chemical Engineering Journal*, 176-177:178–187.

Ragaert, K., Delva, L., and van Geem, K. (2017). Mechanical and chemical recycling of solid plastic waste. *Waste management (New York, N.Y.)*, 69:24–58.

Rahimi, A. and García, J. M. (2017). Chemical recycling of waste plastics for new materials production. *Nature Reviews Chemistry*, 1(6):409.

Ramesh, P. and Vinodh, S. (2020). State of art review on life cycle assessment of polymers. *International Journal of Sustainable Engineering*, 13(6):411–422.

Raoul Meys (2021). Bene94/plastic-supply-chain-model-including-matlab-code-: V1.0.

Reichel, A., Schoenmakere, M., and Gillabel, J. (2016). *Circular economy in Europe: Developing the knowledge base*, volume No. 2/2016 of *EEA report*. Publications Office of the European Union, Luxembourg.

REN21 (2018). Renewables 2018 global status report. paris.

Renouf, M. A., Wegener, M. K., and Nielsen, L. K. (2008). An environmental life cycle assessment comparing Australian sugarcane with US corn and UK sugar beet as producers of sugars for fermentation. *Biomass and Bioenergy*, 32(12):1144–1155.

Ricke, K., Drouet, L., Caldeira, K., and Tavoni, M. (2018). Country-level social cost of carbon. *Nature Climate Change*, 8(10):895–900.

Rieks, M., Bellinghausen, R., Kockmann, N., and Mleczko, L. (2015). Experimental study of methane dry reforming in an electrically heated reactor. *International Journal of Hydrogen Energy*, 40(46):15940–15951.

Rihko-Struckmann, L. K., Peschel, A., Hanke-Rauschenbach, R., and Sundmacher, K. (2010). Assessment of Methanol Synthesis Utilizing Exhaust CO2 for Chemical Storage of Electrical Energy. *Industrial & Engineering Chemistry Research*, 49(21):11073–11078.

Ritter, S. K. (2011). BPA Is Indispensible For Making Plastics. *Chemical and Engineering News*, (Volume 89 Issue 23).

Roberts, D. E. (1950). Heats of polymerization. a summary of published values and their relation to structure. *J. Res. Natl. Bur*, pages 221–232.

Rogelj, J., Popp, A., Calvin, K. V., Luderer, G., Emmerling, J., Gernaat, D., Fujimori, S., Strefler, J., Hasegawa, T., Marangoni, G., Krey, V., Kriegler, E., Riahi, K., van Vuuren, D. P., Doelman, J., Drouet, L., Edmonds, J., Fricko, O., Harmsen, M., Havlík, P., Humpenöder, F., Stehfest, E., and Tavoni, M. (2018). Scenarios towards limiting global mean temperature increase below 1.5 °C. *Nature Climate Change*, 8(4):325–332.

Rönsch, S., Schneider, J., Matthischke, S., Schlüter, M., Götz, M., Lefebvre, J., Prabhakaran, P., and Bajohr, S. (2016). Review on methanation – from fundamentals to current projects. *Fuel*, 166:276–296.

Roser, M. and Ritchie, H. (2021). Future greenhouse gas emissions scenarios. `https://ourworldindata.org/future-emissions#what-trajectory-is-the-world-currently-on`.

Rudolph, N. S., Aumanate, C., and Kiesel, R., editors (2017). *Understanding plastics recycling: Economic, ecological, and technical aspects of plastic waste handling*. Hanser Publishers and Hanser Publications, Cincinnati and Munich.

Saygin, D., Gielen, D. J., Draeck, M., Worrell, E., and Patel, M. K. (2014). Assessment of the technical and economic potentials of biomass use for the production of steam, chemicals and polymers. *Renewable and Sustainable Energy Reviews*, 40:1153–1167.

Schakel, W., Fernández-Dacosta, C., van der Spek, M., and Ramírez, A. (2017). New indicator for comparing the energy performance of CO2 utilization concepts. *Journal of CO2 Utilization*, 22:278–288.

Schlömer, S., Bruckner, T., Fulton, L., Hertwich, E., McKinnon, A., Perczyk, D., Roy, J., Schaeffer, R., Sims, R., Smith, P., and Wiser, R. (2014). Annex iii: Technology-specific cost and performance parameters. In Edenhofer, O., R., Pichs-Madruga, Y., Sokona, E., Farahani, S., Kadner, K., Seyboth, A., Adler, I., Baum, S., and Brunner, P., editors, *Climate Change 2014: Mitigation of Climate Change. Contribution of Working Group III to the Fifth Assessment Report of the Intergovernmental Panel on Climate Change*. Cambridge University Press, Cambridge, United Kingdom and New York, NY, USA.

Schwarz, A. E., Ligthart, T. N., Godoi Bizarro, D., de Wild, P., Vreugdenhil, B., and van Harmelen, T. (2021). Plastic recycling in a circular economy; determining environmental performance through an LCA matrix model approach. *Waste management (New York, N.Y.)*, pages 331–342.

Searchinger, T. D., Hamburg, S. P., Melillo, J., Chameides, W., Havlik, P., Kammen, D. M., Likens, G. E., Lubowski, R. N., Obersteiner, M., Oppenheimer, M., Robertson, G. P., Schlesinger, W. H., and Tilman, G. D. (2009). Climate change. fixing a critical climate accounting error. *Science (New York, N.Y.)*, 326(5952):527–528.

Sethuraj, M. R. and Mathew, N. M. (1992). *Natural rubber: Biology, cultivation, and technology*, volume 23 of *Developments in crop science*. Elsevier, Amsterdam and New York.

Shen, L., Worrell, E., and Patel, M. K. (2010). Open-loop recycling: A LCA case study of PET bottle-to-fibre recycling. *Resources, Conservation and Recycling*, 55(1):34–52.

Simon, N., Raubenheimer, K., Urho, N., Unger, S., Azoulay, D., Farrelly, T., Sousa, J., van Asselt, H., Carlini, G., Sekomo, C., Schulte, M. L., Busch, P.-O., Wienrich, N., and Weiand, L. (2021). A binding global agreement to address the life cycle of plastics. *Science (New York, N.Y.)*, 373(6550):43–47.

Singh, N., Hui, D., Singh, R., Ahuja, I., Feo, L., and Fraternali, F. (2017). Recycling of plastic solid waste: A state of art review and future applications. *Composites Part B: Engineering*, 115:409–422.

Spath, P., Aden, A., Eggeman, T., Ringer, M., Wallace, B., and Jechura, J. (2005). Biomass to Hydrogen Production Detailed Design and Economics Utilizing the Battelle Co-lumbus Laboratory Indirectly-Heated Gasifier. National Renewable Energy Laboratory.

Speight, J. G. (2014). *The Chemistry and Technology of Petroleum, Fifth Edition.* Chemical Industries. Taylor and Francis, Hoboken, 5th ed. edition.

Sphera (2019). *Software-System and Database for Life Cycle Engineering.* Leinfelden-Echterdingen, Germany.

Spierling, S., Knüpffer, E., Behnsen, H., Mudersbach, M., Krieg, H., Springer, S., Albrecht, S., Herrmann, C., and Endres, H.-J. (2018). Bio-based plastics - a review of environmental, social and economic impact assessments. *Journal of Cleaner Production*, 185:476–491.

Stadtherr, M. A. (1978). A systems approach to assessing new petrochemical technology. *Chemical Engineering Science*, 33(7):921–922.

Stadtherr, M. A. and Rudd, D. F. (1976). Systems study of the petrochemical industry. *Chemical Engineering Science*, 31(11):1019–1028.

Stadtherr, M. A. and Rudd, D. F. (1978). Resource use by the petrochemical industry. *Chemical Engineering Science*, 33(7):923–933.

Statistica (2018). U.S. and European natural gas price 1980–2030. Available at https://www.statista.com/statistics/252791/natural-gas-prices. Accessed June 6, 2018.

Sternberg, A. and Bardow, A. (2015). Power-to-what? – environmental assessment of energy storage systems. *Energy & Environmental Science*, 8(2):389–400.

Subhani, M. A., Köhler, B., Gürtler, C., Leitner, W., and Müller, T. E. (2016a). Light-mediated curing of CO2-based unsaturated polyethercarbonates via thiolene click chemistry. *Polym. Chem.*, 7(24):4121–4126.

Subhani, M. A., Köhler, B., Gürtler, C., Leitner, W., and Müller, T. E. (2016b). Transparent Films from CO2-Based Polyunsaturated Poly(ether carbonate)s: A Novel Synthesis Strategy and Fast Curing. *Angewandte Chemie (International ed. in English)*, 55(18):5591–5596.

Sugiyama, H., Fischer, U., Hungerbühler, K., and Hirao, M. (2008). Decision framework for chemical process design including different stages of environmental, health, and safety assessment. *AIChE Journal*, 54(4):1037–1053.

Tanzer, S. E. and Ramírez, A. (2019). When are negative emissions negative emissions? *Energy & Environmental Science*, 12(4):1210–1218.

Thakker, V. and Bakshi, B. R. (2021). Toward sustainable circular economies: A computational framework for assessment and design. *Journal of Cleaner Production*, 295:126353.

Thomson, G. H. (1996). The DIPPR databases. *International Journal of Thermophysics*, 17(1):223–232.

Thonemann, N. (2020). Environmental impacts of CO2-based chemical production: A systematic literature review and meta-analysis. *Applied Energy*, 263:114599.

Tsiropoulos, I., Cok, B., and Patel, M. K. (2013). Energy and greenhouse gas assessment of European glucose production from corn - a multiple allocation approach for a key ingredient of the bio-based economy. *Journal of Cleaner Production*, 43:182–190.

Tsiropoulos, I., Faaij, A., Lundquist, L., Schenker, U., Briois, J. F., and Patel, M. K. (2015). Life cycle impact assessment of bio-based plastics from sugarcane ethanol. *Journal of Cleaner Production*, 90:114–127.

Turner, S. R. (2004). Development of amorphous copolyesters based on 1,4-cyclohexanedimethanol. *Journal of Polymer Science Part A: Polymer Chemistry*, 42(23):5847–5852.

U.S. Environmental Protection Agency (2020). Office of Land and Emergency Management, Office of Resource Conservation and Recovery. Advancing Sustainable Materials Management: 2015 Fact Sheet.

Van-Dal, É. S. and Bouallou, C. (2013). Design and simulation of a methanol production plant from CO2 hydrogenation. *Journal of Cleaner Production*, 57:38–45.

van Ewijk, S., Stegemann, J. A., and Ekins, P. (2018). Global life cycle paper flows, recycling metrics, and material efficiency. *Journal of Industrial Ecology*, 22(4):686–693.

Vinum, M. G., Almind, M. R., Engbaek, J. S., Vendelbo, S. B., Hansen, M. F., Frandsen, C., Bendix, J., and Mortensen, P. M. (2018). Dual-function cobalt-nickel nanoparticles tailored for high-temperature induction-heated steam methane reforming. *Angewandte Chemie (International ed. in English)*, 57(33):10569–10573.

von der Assen, N. and Bardow, A. (2014). Life cycle assessment of polyols for polyurethane production using CO2 as feedstock: insights from an industrial case study. *Green Chem*, 16(6):3272–3280.

von der Assen, N., Jung, J., and Bardow, A. (2013). Life-cycle assessment of carbon dioxide capture and utilization: avoiding the pitfalls. *Energy & Environmental Science*, 6(9):2721.

von der Assen, N., Müller, L. J., Steingrube, A., Voll, P., and Bardow, A. (2016). Selecting CO2 Sources for CO2 Utilization by Environmental-Merit-Order Curves. *Environmental science & technology*, 50(3):1093–1101.

von der Assen, N., Voll, P., Peters, M., and Bardow, A. (2014). Life cycle assessment of CO2 capture and utilization: a tutorial review. *Chemical Society reviews*, 43(23):7982–7994.

von Pfingsten, S., Broll, D. O., von der Assen, N., and Bardow, A. (2017). Second-order analytical uncertainty analysis in life cycle assessment. *Environmental science & technology*, 51(22):13199–13204.

Walker, S. and Rothman, R. (2020). Life cycle assessment of bio-based and fossil-based plastic: A review. *Journal of Cleaner Production*, 261:121158.

Walters, R. N., Hackett, S. M., and Lyon, R. E. (2000). Heats of combustion of high temperature polymers. *Fire and Materials*, 24(5):245–252.

Wang, H., Hodgson, J., Shrestha, T. B., Thapa, P. S., Moore, D., Wu, X., Ikenberry, M., Troyer, D. L., Wang, D., Hohn, K. L., and Bossmann, S. H. (2014a). Carbon dioxide hydrogenation to aromatic hydrocarbons by using an iron/iron oxide nanocatalyst. *Beilstein Journal of Nanotechnology*, 5:760–769.

Wang, Q., Li, X., Li, W., and Feng, J. (2014b). Promoting effect of Fe in oxidative dehydrogenation of ethylbenzene to styrene with CO2 (I) preparation and performance of Ce1−xFexO2 catalyst. *Catalysis Communications*, 50:21–24.

Werpy, T. and Petersen, G. R. (2004). Top value added chemicals from biomass: volume i-results of screening for potential candidates from sugars and synthesis gas. National Renewable Energy Laboratory.

Westhues, S., Idel, J., and Klankermayer, J. (2018). Molecular catalyst systems as key enablers for tailored polyesters and polycarbonate recycling concepts. *Science advances*, 4(8):eaat9669.

Winter, B., Meys, R., and Bardow, A. (2021). Towards aromatics from biomass: Prospective life cycle assessment of bio-based aniline. *Journal of Cleaner Production*, 290:125818.

Wismann, S. T., Engbæk, J. S., Vendelbo, S. B., Bendixen, F. B., Eriksen, W. L., Aasberg-Petersen, K., Frandsen, C., Chorkendorff, I., and Mortensen, P. M. (2019). Electrified methane reforming: A compact approach to greener industrial hydrogen production. *Science (New York, N.Y.)*, 364(6442):756–759.

Wypych, G. (2012). *Handbok of polymers*. ChemTec Publ, Toronto.

Xue, M. and Xu, Z. (2017). Application of life cycle assessment on electronic waste management: A review. *Environmental management*, 59(4):693–707.

Yabe, T., Kamite, Y., Sugiura, K., Ogo, S., and Sekine, Y. (2017). Low-temperature oxidative coupling of methane in an electric field using carbon dioxide over Ca-doped LaAlO 3 perovskite oxide catalysts. *Journal of CO2 Utilization*, 20:156–162.

Yang, H., Zhang, C., Gao, P., Wang, H., Li, X., Zhong, L., Wei, W., and Sun, Y. (2017). A review of the catalytic hydrogenation of carbon dioxide into value-added hydrocarbons. *Catalysis Science & Technology*, 7(20):4580–4598.

You, F., Tao, L., Graziano, D. J., and Snyder, S. W. (2012). Optimal design of sustainable cellulosic biofuel supply chains: Multiobjective optimization coupled with life cycle assessment and input-output analysis. *AIChE Journal*, 58(4):1157–1180.

Young, B., Hottle, T., Hawkins, T., Jamieson, M., Cooney, G., Motazedi, K., and Bergerson, J. (2019). Expansion of the petroleum refinery life cycle inventory model to support characterization of a full suite of commonly tracked impact potentials. *Environmental science & technology*, 53(4):2238–2248.

Zeaiter, J. (2014). A process study on the pyrolysis of waste polyethylene. *Fuel*, 133:276–282.

Zhang, F., Zeng, M., Yappert, R. D., Sun, J., Lee, Y.-H., LaPointe, A. M., Peters, B., Abu-Omar, M. M., and Scott, S. L. (2020). Polyethylene upcycling to long-chain alkylaromatics by tandem hydrogenolysis/aromatization. *Science (New York, N.Y.)*, 370(6515):437–441.

Zhang, Z., Hirose, T., Nishio, S., Morioka, Y., Azuma, N., Ueno, A., Ohkita, H., and Okada, M. (1995). Chemical recycling of waste polystyrene into styrene over solid acids and bases. *Industrial & Engineering Chemistry Research*, 34(12):4514–4519.

Zheng, J., Chen, X., Zhong, X., Li, S., Liu, T., Zhuang, G., Li, X., Deng, S., Mei, D., and Wang, J.-g. (2017). Hierarchical Porous NCCuCo Nitride Nanosheet Networks: Highly Efficient Bifunctional Electrocatalyst for Overall Water Splitting and Selective Electrooxidation of Benzyl Alcohol. *Advanced Functional Materials*, 27(46):1704169.

Zheng, J. and Suh, S. (2019). Strategies to reduce the global carbon footprint of plastics. *Nature Climate Change*, 9(5):374–378.

Zhou, Z., Tang, Y., Chi, Y., Ni, M., and Buekens, A. (2018). Waste-to-energy: A review of life cycle assessment and its extension methods. *Waste management & research : the journal of the International Solid Wastes and Public Cleansing Association, ISWA*, 36(1):3–16.

Zimmerman, J. B., Anastas, P. T., Erythropel, H. C., and Leitner, W. (2020). Designing for a green chemistry future. *Science (New York, N.Y.)*, 367(6476):397–400.

Aachener Beiträge zur Technischen Thermodynamik

ABTT 1
Philip Voll
Automated Optimization-Based Synthesis of Distributed Energy Supply Systems
1. Auflage 2014
ISBN 978-3-86130-474-6

ABTT 2
Johannes Jung
Comparative Life Cycle Assessment of Industrial Multi-Product Processes
1. Auflage 2014
ISBN 978-3-86130-471-5

ABTT 3
Franz Lanzerath
Modellgestützte Entwicklung von Adsorptionswärmepumpen
1. Auflage 2014
ISBN 978-3-86130-472-2

ABTT 4
Thorsten Brands
Einfluss der Gemischzusammensetzung auf die Verbrennung im Diesel- und GCAI-Motor
1. Auflage 2014
ISBN 978-3-95886-006-3

ABTT 5
Dominique Dechambre
Efficient Measurement of Liquid-Liquid Equilibria using Automation and Optimal Experimental Design
1. Auflage 2016
ISBN 978-395886-077-3

ABTT 6
Niklas von der Aßen
From Life-Cycle Assesement towards life-Cycle Design of Carbon Dioxide Capture and Utilization
1. Auflage 2016
ISBN 978-3-95886-080-3

ABTT 7
Matthias Lampe
Integrated Process and Organic Rankine Cycle Working Fluid Design in the Continuous-Molecular Targeting Framework
1. Auflage 2016
ISBN 978-3-95886-086-5

ABTT 8
Thomas Hülser
Optische Untersuchung der Zündvorgänge und deren Auswirkung auf die Verbrennung in PKW-Motoren
1. Auflage 2016
ISBN 978-3-95886-090-2

Aachener Beiträge zur Technischen Thermodynamik

ABTT 9
Malte Döntgen
Reaction Models from Reactive Molecular Dynamics and High-Level Kinetics Predictions
1. Auflage 2016
ISBN 978-3-95886-156-5

ABTT 10
Heike Schreiber
Experiments and Validated Models for Adsorption Thermal Energy Storage in Industrial and Residential Application
1. Auflage 2017
ISBN 978-3-95886-178-7

ABTT 11
André Dirk Sternberg
System-Wide Perspective for Life Cycle Assesment of CO_2-based C1-Chemicals
1. Auflage 2017
ISBN 978-3-95886-193-0

ABTT 12
Uwe Bau
From Dynamic Simulation to Optimal Design and Control of Adsorption Energy Systems
1. Auflage 2018
ISBN 978-3-95886-216-6

ABTT 13
Christian Jens
Modellbasiertes Design von Produkt, Lösungsmittel und Prozess für die Ameisensäure-synthese aus CO_2 und H_2
1. Auflage 2018
ISBN 978-3-95886-231-9

ABTT 14
Jan David Scheffczyk
Integrated Computer-Aided Design of Molecules and Processes using COSMO-RS
1. Auflage 2018
ISBN 978-3-95886-236-4

ABTT 15
Björn Bahl
Optimization-Based Synthesis of Large-Scale Energy Systems by Time-Series Aggregation
1. Auflage 2018
ISBN 978-3-95886-240-1

Aachener Beiträge zur Technischen Thermodynamik

ABTT 16
Bastian Liebergesell
A Milliliter-Scale Setup for the Efficient Characterization of Multicomponent Vapor-Liquid Equilibria Using Raman Spectroscopy
1. Auflage 2018
ISBN 978-3-95886-247-0

ABTT 17
Stefan Wilhelm Graf
A Design Approach for Adsorption Energy Systems Integrating Dynamic Modeling with Small-Scale Experiments
1. Auflage 2018
ISBN 978-3-95886-258-6

ABTT 18
Sebastian Kaminski
Quantum-Mechanics-Based Prediction of SAFT Parameters for Non-Associating and Associating Molecules Containing Carbon, Hydrogen, Oxygen and Nitrogen
1. Auflage 2019
ISBN 978-3-95886-270-8

ABTT 19
Maike Renate Hennen
Decision Support for the Synthesis of Energy Systems by Analysis of the Near-Optimal Solution Space
1. Auflage 2019
ISBN 978-3-95886-277-7

ABTT 20
Peyman Yamin
COSMO-RS-Based Methods for Improved Modelling of Complex Chemical Systems
1. Auflage 2019
ISBN 978-3-95886-288-3

ABTT 21
Meltem Erdogan
Assessement of Adsorbents for Drying by Experiments and Dynamic Simulations
1. Auflage 2019
ISBN 978-3-95886-303-3

ABTT 22
Christian Schulz
SRS/LIF-Messungen zur Charakterisierung rußarmer dieselähnlicher Flammen von alternativen Kraftstoffen und n-Heptan
1. Auflage 2019
ISBN 978-3-91886-310-1

Aachener Beiträge zur Technischen Thermodynamik

ABTT 23
Peter Beumers
Physically-Based Models for the Analysis of Raman Spectra
1. Auflage 2019
ISBN 978-3-95886-319-4

ABTT 24
Arne Kätelhön
Technology Choice Model for Consequential Life Cycle Assessment
1. Auflage 2019
ISBN 978-3-95886-324-8

ABTT 25
Christine Peters
Measurement of Multicomponent Diffusion in Liquids Using Raman Microspectroscopy and Microfluidics
1. Auflage 2020
ISBN 978-3-95886-337-8

ABTT 26
Dinah Elena Hollermann
Reliable and Robust Optimal Design of Sustainable Energy Systems
1. Auflage 2020
ISBN 978-3-95886-346-0

ABTT 27
Thomas Raffius
Laserspektroskopische Analyse von selbstzündenden motorischen Einspritzstrahlen alternativer Biokraftstoffe
1. Auflage 2020
ISBN 978-3-95886-358-3

ABTT 28
Johannes Schilling
Integrated Thermo-Economic Design of Processes and Molecules Using PC-SAFT
1. Auflage 2020
ISBN 978-3-95886-368-2

ABTT 29
Nils Julius Baumgärtner
Optimization of Low-Carbon Energy Systems from Industrial to National Scale
1. Auflage 2020
ISBN 978-3-95886-385-9

ABTT 30
Ludger Wolff
From Model-based Experimental Design and Analysis of Diffusion and Liquid-Liquid Equilibria to Process Applications
1. Auflage 2020
ISBN 978-3-95886-402-3

Aachener Beiträge zur Technischen Thermodynamik

ABTT 31
Andrej Gibelhaus
A Model-based Framework for Optimal Systems Integration of Adsorption Chillers
1. Auflage 2021
ISBN 978-3-95886-406-1

ABTT 32
Jan Seiler
Debottlenecking the Evaporator in Water-Based Adsorption Chillers
1. Auflage 2021
ISBN 978-3-95886-407-8

ABTT 33
Leif Kröger
Prediction of Reaction Rate Constants for the Synthesis of Microgels
1. Auflage 2021
ISBN 978-395886-425-2

ABTT 34
Sarah von Pfingsten
Uncertainty Analysis in Matrix-Based Life Cycle Assessment
1. Auflage 2022
ISBN 978-3-95886-431-3

ABTT 35
Ludger Leenders
Optimization Methods for Integrating Energy and Production Systems
1. Auflage 20022
ISBN 978-3-95886-445-0

ABTT 36
Leonard Müller
Harmonized Life Cycle Assessment of Technologies for Carbon Capture and Utilization
1. Auflage 2022
ISBN 978-3-95886-434-4

ABTT 37
Fritz Röben
Decarbonization of Copper Production by Optimal Demand Response
and Power-to-Hydrogen
1. Auflage 2022
ISBN 978-3-95886-458-0

ABTT 38
Johanna Kleinekorte
Predictive Life Cycle Assessment for Chemical Processes using Machine Learing
1. Auflage 2022
ISBN 978-3-95886-461-0

Aachener Beiträge zur Technischen Thermodynamik

ABTT 39
Raoul Meys
Designing Pathways for Net-Zero Greenhouse Gas Emission Plastics
with Life Cycle Optimization
1. Auflage 2022
ISBN 978-3-95886-463-4